MEP 016B 3KW Generator
Intermediate Maintenance Manual
TM 5-6115-615-34

Generator Set
Skid Mounted, Tactical Quiet

edited by
Brian Greul

The MEP series of Military Generators are rugged, durable and incorporate proven diesel engine technology. This book is the intermediate maintenance manual. It is being republished to assist enthusiasts, restorers, and aftermarket owners who use or wish to use these generators outside of military use.

An 8.5x11 3 hole punched loose leaf copy may be purchased for your 3 ring binder. Email books@ocotillopress.com for current information.

Should you have suggestions or feedback on ways to improve this book please send email to Books@OcotilloPress.com

Edited 2021 Ocotillo Press
ISBN 978-1-954285-25-5

Printed in the United States of America

Ocotillo Press
Houston, TX 77017
Books@OcotilloPress.com

Disclaimer: The user of this book is responsible for following safe and lawful practices at all times. The publisher assumes no responsibility for the use of the content of this book. The publisher has made an effort to ensure that the text is complete and properly typeset, however omissions, errors, and other issues may exist that the publisher is unaware of.

MARINE CORPS TECHNICAL MANUAL
ARMY TECHNICAL MANUAL
NAVY PUBLICATION NAVFAC P-8-646-34
AIR FORCE TECHNICAL ORDER

TM 05926B/06509B-34/3 TM

TO 35C2-3-386-32

TECHNICAL MANUAL

INTERMEDIATE (FIELD) (DIRECT AND GENERAL SUPPORT)
MAINTENANCE MANUAL

GENERATOR SET, DIESEL ENGINE DRIVEN, TACTICAL,
SKID MOUNTED, 3 KW, 3 PHASE 120/208 AND
SINGLE PHASE 120/240 VOLTS AC AND 28 VOLTS DC

DOD MODEL	CLASS	MODE	NSN
MEP-016B UTILITY 60 HZ 6115-01-150-4140			
MEP-021B	UTILITY	400 HZ	6115-01-151-8126
MEP-026B	UTILITY	28 VDC	6115-01-150-0367

PUBLISHED UNDER THE AUTHORITY OF HEADQUARTERS U.S. MARINE CORPS,
THE DEPARTMENTS OF THE ARMY, AIR FORCE, AND NAVY

JULY 1987

PCN 184 059270 00

MARINE CORPS TECHNICAL MANUAL	TM 05926B/06509B-34/3
DEPARTMENT OF THE ARMY TECHNICAL MANUAL	TM 5-6115-615-34
DEPARTMENT OF THE NAVY PUBLICATION	NAVFAC P-8-646-34
DEPARTMENT OF THE AIR FORCE TECHNICAL ORDER	TO 35C2-3-386-32

HEADQUARTERS U.S. MARINE CORPS, DEPARTMENTS OF
THE ARMY, NAVY, AND AIR FORCE
WASHINGTON, D.C. (July 1987)

INTERMEDIATE (FIELD) (DIRECT AND GENERAL SUPPORT)
MAINTENANCE MANUAL

GENERATOR SET, DIESEL ENGINE DRIVEN, TACTICAL, SKID MOUNTED, 3 KW, 3 PHASE 120/208 AND SINGLE PHASE 120/240 VOLTS AC AND 28 VOLTS DC

DOD MODEL	CLASS	MODE	N S N
MEP-016B	UTILITY	60 HZ	6115-01-150-4140
MEP-021B	UTILITY	400 HZ	6115-01-151-8126
MEP-026B	UTILITY 28 VDC	6115-01-150-0367	

Marine Corps TM 05926B/06509B-34/3 Army TM
5-6115-615-34
Navy NAVFAC P-8-646-34 Air
Force TO 35C2-3-386-32

DEPARTMENT OF THE NAVY Headquarters, U.S. Marine Corps
Washington, D.C. 20380-0001

31 July 1987

1. This Manual is effective upon receipt and contains intermediate
 (Field) (Direct and General Support) maintenance instructions for Generator Set, Engine Driven, Tactical, Skid Mounted, 3 KW, DOD Models MEP-016B, 60 HZ, NSN 6115-01-150-4140, MEP-021B,
400 HZ,
NSN 6115-01-151-8126, MEP-026B, 28 VDC, NSN 6115-01-150-0367.

2.   Notice of discrepancies or suggested changes: refer to paragraph 1-4
titled, Reporting of Errors, of this Manual for applicable Services' form number and forwarding address.

BY DIRECTION OF THE COMMANDANT OF THE MARINE CORPS

OFFICIAL:

J. G. O'NEILL
Head, Materiel Acquisition Support Branch
Materiel Division
Installations and Logistics Department

CARL E. VUONO
General, United States Army Chief of
Staff

R. L. DILWORTH
Brigadier General, United States Army
The Adjutant General

B. F. MONTOYA, RADM, CEC USN

LARRY D. WELSH, General USAF,
Chief of Staff

ALFRED G. HANSEN, General, USAF
Commander, Air Force Logistics Command

DISTRIBUTION:AGB/L77/L82

Copy to: 7000161(2)

LIST OF EFFECTIVE PAGES

INSERT LATEST CHANGED PAGES. DESTROY SUPERSEDED PAGES.

NOTE: The portion of the text affected by the changes is indicated by a vertical line in the outer margins of the page. Changes to Illustrations are indicated by miniature pointing hands. Changes to wiring diagrams are indicated by shaded areas.

Date of issue for original and changed pages are:
Original . . . 0 . . . July 1987

TOTAL NUMBER OF PAGES IN THIS PUBLICATION IS 198 CONSISTING OF THE FOLLOWING:

Page No.	* Change No.
Cover	0
Blank	0 0
Title Block	0 0
Blank	0
A	0 0
Blank. i	0
thru x 1-1	0
thru 1-18 . . . . 2-1 thru	0 0
2-29 . . . . Blank	0
. . . . 3-1 thru 3-10	0
. . . . 4-1 and 4-2	0 0
. . . . . . 5-1 thru 5-33 . . . . ,, .	0
. Blank 6-1	0
thru 6-12 . . . . 7-1 thru	0
7-60 . . . . 8-1	0 0
. . . . Blank	0
. . . . 9-1 thru 9-3	0
. . . . . Blank	0
10-1 thru 10-4 . . . . A-1 . .	
. . . .	
Blank	
Index-1 thru Index-3	
Blank	

* Zero in this column indicates an original page.

MARINE CORPS TM 05926B/06509B-34
ARMY   TM 5-6115-615-34
NAVY   NAVFAC P-8-646-34
AIR FORCE   TO 35C2-3-386-32

TABLE OF CONTENTS

Chapter/Section/Paragraph

CHAPTER 1.   INTRODUCTION 1-1
Section I.   General 1-1
   1-1.   Scope 1-1
   1-2.   Limited Applicability 1-1
   1-3.   Maintenance Forms and Records 1-1
   1-4.   Reporting of Errors 1-1
   1-5.   Levels of Maintenance Accomplishment 1-2
   1-6.   Destruction of Army Materiel to Prevent Enemy Use 1-2
   1-7.   Administrative Storage 1-2
   1-8.   Preparation for Shipment and Storage 1-2
   1-9.   Reporting Equipment Improvement Recommendations, MC 1-2
   1-10.   Reporting Equipment Improvement Recommendations, A 1-2
Section II. Description and Tabulated Data 1-3
   1-11.   Description 1-3
   1-12.   Tabulated Data 1-3
   1-13.   Differences Between Models 1-18

CHAPTER 2.   GENERAL MAINTENANCE INSTRUCTIONS 2-1
Section I.   Repair Parts, Special Tools, Test, Measurement, and
                Diagnostic Equipment (TMDE), and Support Equipment 2-1
   2-1.   Repair Parts 2-1
   2-2.   Tools and Equipment 2-1
   2-3.   Fabricated Tools And Equipment 2-3
Section II. Troubleshooting 2-6
   2-4.   General 2-6
   2-5.   Malfunctions Not Corrected By Use of
                Troubleshooting Table 2-6
Section III. General Maintenance 2-14
   2-6.   General Maintenance 2-14
   2-7.   General Maintenance Requirements 2-14
Section IV. Removal and Installation of Major Components 2-16
   2-8.   Control Box 2-16
   2-9.   Engine 2-21
   2-10.   Generator 2-25

CHAPTER 3.   MAINTENANCE OF THE FRAME 3-1
   3-1.   General 3-1
   3-2.   Frame 3-1
   3-3.   Skid Base 3-8

CHAPTER 4.   MAINTENANCE OF THE DC ELECTRICAL
             AND CONTROL SYSTEM 4-1
   4-1.   General 4-1
   4-2.   Battery Repair 4-1

CHAPTER 5.   MAINTENANCE OF THE ELECTRICAL POWER
             GENERATION AND CONTROL SYSTEM 5-1
   5-1.   General 5-1
   5-2.   Generator Assembly 5-1

MARINE CORPS TM 05926B/06509B-34
ARMY   TM 5-6115-615-34
NAVY   NAVFAC P-8-646-34
AIR FORCE   TO 35C2-3-386-32

TABLE OF CONTENTS - continued

Chapter/Section/Paragraph   Page

5-3.   Generator Fan 5-6
5-4.   Rotor 5-7
5-5.   Rotating Rectifiers (Diodes) 5-12
5-6.   Stator Testing 5-13
5-7.   Excitor Stator Testing   5-16
5-8.   Generator Bearing   5-17
5-9.   Voltage Regulator . . . . 5-18 . . . .
5-10.   Current Transformer MEP-D16B/MEP021B (60/400 Hz)
           Sets Only   5-32

CHAPTER 6.   MAINTENANCE OF THE FUEL SYSTEM   6-1
6-1.   General 6-1
6-2.   Fuel Tank   6-1
6-3.   Fuel Injection Pump   6-3
6-4.   Fuel Injector   6-7

CHAPTER 7.   MAINTENANCE OF THE ENGINE   7-1
7-1.   General   7-1
7-2.   Engine Assembly   7-1
7-3.   Oil Pan 7-8
7-4.   Starter Assembly   7-10
7-5.   Starter Solenoid 7-19
7-6.   Cylinder Head   7-20
7-7.   Valves 7-25
7-8.   Rocker Arms and Push Rods   7-27
7-9.   Lifters and Push Rod Tubes   7-30
7-10.   Flywheel and Engine Fan 7-32
7-11.   Battery Charger Stator and Air Scroll Back Plate   7-35
7-12.   Gear Cover and Seal   7-38
7-13.   Cylinder and Piston   7-40
7-14.   Oil Pump   7-44
7-15.   Connecting Rod   7-45
7-16.   Camshaft   7-47
7-17.   Crankshaft and Flywheel Housing   7-49
7-18.   Oil Filter Adapter 7-53
7-19.   Crankcase 7-55
7-20.   . . . .
    Oil Cooler 7-59

CHAPTER 8.   MAINTENANCE OF THE ENGINE CONTROLS AND INSTRUMENTS   8-1
8-1.   General 8-1
8-2.   Engine Control Circuit Board   8-1

TABLE OF CONTENTS - continued

CHAPTER 9.   MAINTENANCE OF GENERATOR CONTROLS AND INSTRUMENTS 9-1
        9-1.   General 9-1
        9-2.   Control Box Assembly   9-1
        9-3.   Frequency Meter 9-1
        9-4.   Frequency Transducer   9-1

CHAPTER 10.   GENERATOR SET TEST AND INSPECTION
               AFTER REPAIR OR OVERHAUL   10-1
Section I.   General Requirements   10-1
       10-1.   General 10-1
Section II.   Inspection 10-1
       10-2.   Generator Set 10-1
       10-3.   Engine   10-1
       10-4.   Wiring Harnesses 10-1
Section III.   Operational Test   10-1
       10-5.   Operational Tests 10-1

APPENDIX A REFERENCES A-1

INDEX Index-1

MARINE CORPS   TM 05926B/06509B-34
ARMY   TM 5-6115-615-34
NAVY   NAVFAC P-8-646-34
AIR FORCE   TO 35C2-3-386-32

LIST OF ILLUSTRATIONS

Figure	Title	Page
1-1	Cylinder Head Torquing Sequence	1-18
2-1	Control Box Removal and Installation	2-17
2-2 2-3	Generator Wiring MEP-026B	2-18
2-4 2-5	Generator Wiring MEP-016B and MEP-021B	2-19
2-6	Control Box Harness	2-20
	Engine Removal	2-21
3-1	Generator Removal and Installation	2-26
3-2		
	Frame Removal	3-2
5-1	Skid Base	3-9
5-2		
5-3	Generator Testing- MEP-016B and MEP-021B	5-2
5-4	Generator Testing - MEP-026B	5-3
5-5	Generator Disassembly and Assembly	5-5
5-6	Generator Fan	5-7
5-7	Rotor	5-7
5-8	Rotor Insulation Test	5-9
5-9	Main Rotor Winding Resistance Check	5-10
5-10	Exciter Rotor Winding Resistance Check	5-11
5-11	Rotating Rectifiers	5-12
5-12	Stator Winding Insulation Test	5-14
5-13	Generator Wiring Schematics	5-16
5-14	Generator Bearing	5-17
5-15	Voltage Regulator Testing	5-18
5-16	Q1 and Q2 Transistor Pin Out Locations	5-19
5-17	Voltage Regulator Schematic	5-23
5-18 5-19	Voltage Regulator Adjustment Schematic	5-26
5-20	Waveforms	5-26
5-21	Voltage Regulator MEP-016B and MEP-021B (60/400 Hz Sets)	5-29
	Voltage Regulator MEP-026B (28VDC Set)	5-30
6-1	Standard Voltage Regulator Jumper Wire Placement	5-33
6-2	Current Transformer	5-34
6-3		
6-4	Fuel Tank	6-2
	Fuel Injection Pump	6-4
6-5 6-6	External Governor Linkage	6-7
7-1	Fuel Injector Removal and Installation	6-8
7-2	Fuel Injector Testing	6-9
7-3	Fuel Injector	6-11
7-4 7-5		
7-6	Low Fuel Shutdown Solenoid Adjustment	7-5
	Governor Linkage Adjustment	7-6
	Governor Adjustment	7-7
	Oil Pan	7-9
	Starter Disassembly/Assembly	7-11
	Testing Armature For Grounds or Shorts	7-12

LIST OF ILLUSTRATIONS - continued.

Figure		Title	Page
7-7		Testing Armature For Open Circuits	7-13
7-8		Solenoid Testing	7-14
7-9		Brush Wear Limit	7-15
7-10		Overrunning Clutch	7-16
7-11		Pinion Gear Installation	7-17
7-12		Adjusting Pinion Shaft End Play	7-17
7-13		Lever Installation	7-18
7-14		Pinion Gap Adjustment	7-19
7-15		Starter Solenoid	7-20
7-16		Cylinder Head	7-21
7-17		Valves	7-25
7-18	7-19	Rocker Arms and Push Rods	7-28
7-20		Piston at Top Dead Center to Set Engine Timing	7-29
7-21		Adjusting Valve Clearance	7-30
7-22		Push Rod Tubes and Lifters	7-31
7-23		Flywheel and Engine Fan	7-33
7-24		Fan and Drive Adapter Alinement Tabs	7-34
7-25		Battery Charger Stator Testing	7-35
7-26		Battery Charger Stator	7-36
7-27		Gear Cover and Seal	7-39
7-28	7-29	Cylinder and Piston	7-41
7-30	7-31	Piston Ring Positioning	7-43
7-32	7-33	Oil Pump	7-44
7-34	7-35	Connecting Rod	7-46
		Camshaft	7-48
		Crankshaft and Generator Adapter	7-50
9-1		Oil Filter Adapter	7-54
		Crankcase	7-56
		Engine Oil Cooler	7-59
		Frequency Transducer Testing	9-2

LIST OF TABLES

Table   **Title**   **Page**

1-1   Repair and Replacement Standards 1-7
1-2   Critical Torque Values 1-16
1-3   Standard Torque Values   1-18

2-1   Special Tools, Test and Support Equipment 2-1
2-2   Consumable Operating and Maintenance Supplies 2-4
2-3   Troubleshooting 2-6

5-1   Q2 Test Chart   5-19
5-2   Q1 Test Chart   5-21
5-3   Test Conditions 5-27

7-1   Compression Test Faults   7-2

10-1   Test Schedule (MEP-016B/60 Hz)   10-1
10-2   Test Schedule (MEP-021B/400 Hz)   10-3
10-3   Test Schedule (MEP-026B/28 VDC)   10-4

SAFETY SUMMARY

The following are general safety precautions that are not related to any specific procedures and therefore do not appear elsewhere in this publication.These are recommended precautions that personnel must understand and apply during many phases of operation and maintenance.

KEEP AWAY FROM LIVE CIRCUITS

Operating personnel must at all times observe all safety regulations.DO not replace components or make adjustments to this equipment with the high voltage supply turned on.Under certain conditions, dangerous potentials may exist when the power control is in the off position, due to charges retained by capacitors. To avoid casualties, always remove power and discharge and ground a circuit before touching it, and always remove rings, watches and other jewelry before servicing this equipment.

DO NOT SERVICE OR ADJUST ALONE

Under no circumstances should any person reach into this equipment for the purpose of servicing or adjusting this equipment except in the presence of someone who is capable of rendering aid.

RESUSCITATION

Personnel working with or near high voltages should be familiar with modern methods of resuscitation.   Such information may be obtained from the Bureau of Medicine and Surgery.

SECURE LOOSE CLOTHING

Personnel working on this equipment should secure all loose fitting clothing to prevent clothing from catching in moving parts.

KEEP COMPRESSED AIR AWAY FROM SKIN

Personnel using compressed air should not exceed 15 psi nozzle pressure when drying parts, and should not direct compressed air toward skin.Personal injury could result.

OPERATE EQUIPMENT IN A WELL VENTILATED AREA

Exhaust discharge contains noxious and deadly fumes. DO not operate this equipment in an enclosed area unless exhaust discharge is properly vented to the outside.When using cleaning solvents, clean parts in a well ventilated area and avoid inhalation of solvent fumes.

WEAR EAR PROTECTION
The noise level of this generator set can cause hearing damage.   To avoid hearing
damage, always wear ear protectors, as recommend by the medical or safety officer,
when operating near this equipment.

USE CAUTION WHEN WORKING AROUND FLAMMABLES
 Do not smoke, use open flame or use excessive heat in the vicinity of this equipment
when refueling, working around the battery or working with flammable cleaning
solvents.   Doing so could cause could result in severe personal

an explosion which injury or
death.

The following   warnings appear in the text and are repeated here for emphasis.

WARNING

DRY CLEANING SOLVENT, P-D-680 or P-S-681, is used to clean parts and is
potentially dangerous to personnel and property.Avoid repeated and prolonged skin
contact.   DO NOT use near open flame or excessive heat. Flash point of solvent
is 100° to 138°F (38° to 60°C).
Paragraph 2-7.
Connect battery cables last when installing frame. The high current output of the
DC electrical system can cause arcing and/or burns if a short circuit occurs. Paragraph
3-2.

WARNING

Disconnect battery cables before removing frame. The high current output of the
DC electrical system can cause arcing and/or burns if a short circuit occurs. Paragraph
3-2.

WARNING

```
—— WARNING ——
```

Observe safety regulations. The voltages used in this test are dangerous to human life. Contact with the leads or windings under test can cause severe,
possibly fatal, shock.Arrange the high voltage leads so that they are not in a position to be accidentally touched. Keep clear of all energized parts.Always reduce the test voltage to zero and ground the winding under test before making any mechanical or electrical adjustments on the equipment. When grounding out windings which have been tested, always connect the connection wire to ground first and then to the winding. Never perform this test without at least one other person assisting. Generator frame shall be securely grounded.Paragraphs 5-4 and 5-6.
Turn off 115 VAC power source before attempting any inspection or repair.Paragraph 5-9.
Wear protective clothing and face shield when opening fuel injection line. Fuel under high pressure may be trapped in the fuel injection line. Opening fuel injection line can cause a high pressure stream of fuel to be released which can cause severe

```
—— WARNING ——
```

personal injury. Paragraphs 6-4 and 6-7.

```
—— WARNING ——
```

Do not allow nozzle to spray against skin. Fuel under nozzle pressure can penetrate flesh infection.Paragraph 6-4.

```
—— WARNING ——
```

and cause a serious

```
—— WARNING ——
```

Be sure to turn growler off between will be produced.   intervals or voltage

```
—— WARNING ——
```

Do not use a punch, prybar, or chisel to remove valve seat.   Seat is made of a hardened material which may shatter and cause personal injury.Paragraph 7-6.

WARNING

Wear protective gloves when handling dry ice and cooled valve seat.Personal injury may result if the seat or ice comes in contact with unprotected skin. Paragraph 7-6. Hot oil and heated crankshaft drive gear can severely burn exposed areas of skin. Wear insulated gloves and protective clothing to protect against burns. Paragraph 7-17. Before repairing battery posts, place battery on a work bench under an exhaust hood to protect personnel from lead and /or acid fumes.Paragraph 4-2.

WARNING

Always wear safety goggles and gloves when repairing battery posts. Paragraph 4-2.

WARNING

WARNING

CHAPTER 1

INTRODUC-

Section I. GENERAL

1-1. SCOPE.This manual is for your use in maintaining the 3KW DED Generator Set, Type I (Tactical), Class 2 (Utility), skid mounted Models MEP-016B, MEP-021B and MEP-026B.

1-1.1. Contents.This manual covers Intermediate (Field) (Direct and General Support Maintenance).   It contains sections for general maintenance, troubleshooting, removal and replacement of components, and maintenance of individual components of the unit.

1-1.2. Appendices. Appendix A contains a list of reference publications applicable to this manual.

1-2. **LIMITED APPLICABILITY.**

Some portions of this publication are not applicable to all services.   These portions are prefixed to indicate the services to which they pertain: (A) for Army,
 (AF) for Air Force, (N) for Navy and (MC) for Marine Corps. Portions not prefixed are applicable to all services.

1-3. **MAINTENANCE FORMS AND RECORDS.**

1-3.1. (MC) Maintenance forms and records used by Marine Corps personnel are prescribed in TM 4700-15/1.

1-3.2. (A) Maintenance forms and records used by Army PAM 738-750.   personnel are prescribed in DA

1-3.3. (AF) Maintenance forms and records used by Air in AFM-66-1 and the applicable 00-20 Series Technical   Force personnel are prescribed Orders.

1-3.4. (N) Navy users should applicable   refer to their service peculiar directives to determine and records to be used.
maintenance forms

1-4. **REPORTING OF ERRORS.**

                              and recommendations for improvement of this publication

by the individual user is encouraged. Reports should be submitted as follows:

1-4.1. (MC) By NAVMC Form 10772 directly to: Commanding General, Marine Corps Logistic Base (Code 850), Albany, GA 31704-5000.

1-4.2. (A) DA Form 2028 directly to: Commander, US Army Troop Support Command, ATTN: AMSTR-MCTS Goodfellow Boulevard, St. Louis, MO 63120-1798.

1-4.3. (AF) AFTO Form 22 directly to:Commander, Sacramento Air Logistics Center, ATTN: MMEDT, McClellan Air Force Base, CA 95652 in accordance with TO-00-5-l.

1-4.4. (N) By letter directly to: Commanding Officer, U.S. Navy, Ships Parts Control Center, ATTN:Code 783, Mechanics-burg, PA 17055.

1-5. LEVELS OF MAINTENANCE ACCOMPLISHMENT.

1-5.1. (MC)Marine Corps users shall refer to the Repair Parts List.

1-5.2. (A) Army users shall refer to the Maintenance Allocation Chart (MAC) for tasks and levels of maintenance to be performed.

1-5.3. (AF) Air Force users shall accomplish maintenance at the user level consistent with their capability in accordance with policies established in accordance with AFM 66-1.

1-5.4. (N) Navy users shall determine their maintenance levels in accordance with their service directives.

1-6. (MC, A) DESTRUCTION OF ARMY MATERIEL TO PREVENTENEMY USE.Demolition of
materiel to prevent enemy use shall be in accordance with the requirement of TM-750-244-3 (Procedures for Destruction of Equip-ment to Prevent Enemy Use for U.S. Army).

1-7.   ADMINISTRATIVE STORAGE.

1-7.1.(MC, N) Refer to individual service directives for   requirements.

1-7.2. (A) Refer to TM 740-90-1 (Administrative Storage).

1-7.3. (AF) Refer to TO 35-1-4 (Processing and Inspection Equipment). of Aerospace Ground

1-8. PREPARATION FOR SHIPMENT AND STORAGE.

1-8.1.(MC Refer to MCO P4450.7.

1-8.2.(A) Refer to TB 740-97-2 and TM 740-90-1.

1-8.3. (AF) Refer to TO 35-1-4 for end item generator sets and TO 38-1-5 for installed engine.

1-8.4. (N) Refer to individual service directives for requirements.

1-9. (MC) REPORTING EQUIPMENT IMPROVEMENTRECOMMENDATIONS (EIR).Submit Quality Assurance
Report on standard form 368 in accordance with MCO 4855.10.

1-10. (A) REPORTING EQUIPMENT IMPROVEMENTRECOMMENDATIONS (EIR). EIR will be prepared using
DA form 2407, Maintenance are provided in DA PAM 738-750, The Army should be mailed directly to:U.S. Army 4300 Goodfellow Boulevard, St. Louis, MO directly to you. Request.   Instructions for preparing EIR's Maintenance Management System.EIR's Troop Support Command, ATTN: AMSTR-QX, 63120-1798.A reply will be furnished

MARINE CORPS TM 05926B/06509B-34
ARMY   TM 5-6115-615-34
NAVY   NAVFAC P-8-646-34
AIR FORCE   TO 35C2-3-386-32

Section II.DESCRIPTION and TABULATED DATA

1-11. DESCRIPTION.

A general description of the diesel engine generator sets and information pertaining to the identification plates are contained in the Operator/Crew and Organizational Maintenance Manual for this unit.Detailed descriptions of the components of the diesel engine generator sets are of this manual.
provided in the applicable maintenance paragraphs

1-12. TABULATED DATA.

1-12.1. General. This paragraph Intermediate (Field) (Direct and
contains all maintenance data pertinent to General Support).

a. Engine, Generator Sets.
Manual.
Refer to Operator and Organizational Maintenance

b. Main Generator. Exciter
field voltage and current versus load as follows:

(1)
Model MEP-016B:0.38 amp. max. at no load, 0.90 amp. max. at rated load Model MEP-021B:0.60 amp. max.at
(2)

no load, 0.96 amp. max. at rated load

(1)

(3) Model MEP-026B:0.25 amp. max. at no load, 0.77 amp. max. at rated load c. Generator

Stator.
Model MEP-016B:

DOD Drawing Number . . . . 13213E4079 (Rev. P) . . . .
Number of Slots 27
Pitch of Coil.   1-10
Coil Grouping 5-4-5/4-5-4

(2) Coil Sides Per Slot   2
Turns Per Coil Group 9
Conductor . . . . 2 AWG No. 16 Resistance Between Lead Pairs at 25°C (77°F) . .
337 ohms ± 10 percent
Model MEP-021B:

DOD Drawing Number. . . . . 13213E4148 (Rev. P)
Number of Slots.   60
Pitch of Coil.   1-4
Coil Group 212/121/121/212/112/121/211
Coils Per Group 2
. . . .

MARINE CORPS TM 05926B/06509B-34
ARMY   TM 5-6115-615-34
NAVY   NAVFAC P-8-646-34
AIR FORCE   TO 35C2-3-386-32

(3) Model MEP-026B:

DOD Drawing Number .... 13213E4108 (Rev. K)

Number of Slots 27

Pitch of Coil 1-10

Coil Groups 5-4-3/4-5-4

Coils Per Group 2

Turns Per Coil Group 3/2/3/2/3/2/3/2/3 (3)

Conductor .... 6- No. 16 AWG

Resistance Between Lead Pairs at 25°C
(77°F) .... 0.02 ohms ± 10 percent

d.   Generator Rotor.

(1)   Model MEP-016B:

DOD Drawing Number 13213E4222

Number of Poles 2

Turns Per Pole 1,140

Conductor .... 18.5 AWG Connection .... Series

Resistance of Connected Poles at 25°C
(77°F) .... 10.6 ohms ± 10 percent

Model MEP-021B

(2)   DOD Drawing Number 13213E4170

Number of Poles 14

Turns Per Pole 95

Conductor .... 3 AWG No. 21 Connection ....

Series Resistance of Connected Poles at 25°C
(77°F) .... 6.1 ohms ± 10 percent

Model MEP-026B

DOD Drawing Number 13213E4222

(3)   Number of Poles 2

Turns per Pole 1,140

Conductor .... 18.5 AWG Connection .... Series

Resistance of Connected Poles at 25°C
(77°F) .... 10.6 ohms ± 10 percent

e.   Exciter Stator.

(1) Model MEP-016B:

DOD Drawing Number D13213E4080
Number of Poles 6
Turns Per Coil 385
Conductor .... No. 23 AWG Connection .... Series
Resistance at 25°C (77°F) .... 22.6 ohms ± 10 percent

(2) Model MEP-021B:

DOD Drawing Number D13213E4080
Number of Poles 6
Turns Per Coil 385
Conductor .... No. 23 AWG Connection .... Series
Resistance at 25°C (77°F) .... 22.6 ohms ± 10 percent

(3) Model MEP-026B:

DOD Drawing Number D13213E4080
Number of Poles 6
Turns Per Coil 385
Conductor .... No. 23 AWG Connection .... Series
Resistance at 25°C (77°F) .... 22.6 ohms ± 10 percent

f.   Exciter Rotor.

(1) Model MEP-016B:

DOD Drawing Number 13213E4074
Number of Slots 45
Pitch of Coil 1-7
Coil Groups 2-3-2/3-2-3
Coils Per Group 2
Turns Per Coil 11
Conductor .... AWG NO. 21
Resistance Between Leads at 25°C (77°F) ...... 2.4 ohms ± 10 percent

(2) Model MEP-021B:

DOD Drawing Number 13213E4074
Number of Slots 45
Pitch of Coil 1-7
Coil Grouping 2-3-2/3-2-3
Coils Sides Per Slot 2
Turns Per Coil 11
Conductor .... AWG No. 21
Resistance Between Leads at 25°C (77°F) ...... 2.4 ohms ± 10 percent

(3)Model MEP-026B:

DOD Drawing Number 13213E4074
Number of Slots 45
Pitch of Coil 1-7
Coil Groups 2-3-2/3-2-3
Coils Per Group 2
Turns Per Coil 11
Conductor . . . . AWG No. 21
Resistance Between Leads at 25°C (77°F) . . . . . . 2.4 ohms ± 10 percent

g. Circuit Breaker.

(1) Model MEP-016B:

DOD Drawing Number 13208E5838
Type AM333M66
Voltage . . . . 250 AC Frequency
400 or 60 Hz
Interrupting Capacity . . . . 1000 AMPS Relay Coil 60 Hz Trip Poles 1
& 2, 54 to 70 AMPS
Trip . . . . Pole 3, 45 to 54 AMPS Relay Coil . . . . 400 Hz
Trip . . . . Poles 1 & 2, 70 to 90 AMPS Trip   Pole 3, 42 to 54 AMPS
Temperature Range . . . . -65 to 175 degrees Fahrenheit

h. Repair and Replacement Standards. Refer to Table 1-1 for repair and
replacement standards.

Table 1-1. Repair and Replacement Standards.

| Component | Manufacturer Dimensions and Tolerances | | Maximum Allowable Wear Limit |
| | Minimum | Maximum | |
	Inches (mm)		
CRANK-			
Main bearing bore (without bearing)	3.1845 (80.886)	3.1854 (80.910) (80.910)	3.1854
Main bearing bore (bearing installed)	2.9947 (76.067)	2.9967 (76.117)	Refer to main journal bearing clearance

NOTE

Main bearings are available in four undersizes: 0.01
(0.25 mm), 0.02 (0.5 mm), 0.03 (0.75 mm) and 0.04 (1.00
mm) .

Component	Minimum	Maximum	Maximum Allowable Wear Limit
Camshaft bearing bore- front (without bearing	1.9354 (49.158	1.9363 (49.182)	1.9363 (49.182)
Camshaft bearing bore - rear (without bearing)	1.1079 (28.142)	1.1089 (28.166)	1.1089 (28.166)
Camshaft bearing bore - front (bearing installed)	1.8129 (46.049)	1.8149 (46.099)	Refer to camshaft bearing clearance
Camshaft bearing bore - rear (bearing installed)	0.9858 (25.040)	0.9878 (25.090)	Refer to camshaft bearing clearance
Valve tappet bore	0.986 (25.044)	0.9872 (25.074)	Refer to valve tappet clearance in bore
Rear bearing housing main bearing bore (without bearing)	3.1837 (80.866)	3.1846 (80.890)	3.1846 (80.890)
Rear bearing housing main bearing bore (bearing installed)	2.9943 (76.055)	2.9962 (76.104)	Refer to main bearing journal clearance

Table 1-1. Repair and Replacement Standards - continued.

| Component | Manufacturer Dimensions and Tolerances | | Maximum Allowable |
| | Minimum | Maximum | |
	Inches (mm) Wear Limit		
Governor actuator bushing (reamed in place)	0.311 (7.900)	0.312 (7.925)	0.317 bore (8.052)
Governor actuator shaft	0.3093 (7.856)	0.3103 (7.882) (7.831)	0.3083

CRANK-

Component	Minimum Inches (mm)	Maximum Wear Limit	Maximum Allowable
Main bearing journal diameter	2.9917 (75.990)	2.9925 (76.010)	Refer to main bearing journal clearance
Main bearing journal out of round		0.0003 (0.007)	0.0004 (0.010)
Main bearing journal taper		0.0002 (0.005)	0.0003 (0.007)
Connecting rod journal diameter	2.361 (59.99)	2.362 (60.01)	Refer to connecting rod to crankshaft clearance (0.006) (0.15)
Crankshaft (Front) main bearing clearance	0.0022 (0.057)	0.0050 (0.127)	0.0055 (0.14)
Crankshaft (Rear) main bearing clearance	0.0017 (0.045)	0.0045 (0.114)	0.016 (0.4)
Crankshaft thrust clearance	0.008 (0.20)	0.014 (0.35)	0.157 (4.00)
Crankshaft thrust washer thickness	0.157 (4.000)	0.159 (4.050)	
Seal surface diameter (front)	1.967 (49.98)	1.969 (50.01)	

Table 1-1. Repair and Replacement Standards - continued.

Component	Manufacturer Dimensions and Tolerances		Maximum Allowable Wear Limit
	Minimum	Maximum	
	Inches (mm)		

CONNECTING

Component	Minimum	Maximum	Max Wear
Side clearance	0.004 (0.1)	0.016 (0.4) (0.4)	0.016
Length (center to center)	6.1407 (155.975)	6.1427 (156.025) (156.025)	6.1427
Bearing bore (less bearing and rod tightened to the proper torque)	2.5141 (63.860)	2.5147 (63.875)	2.5147 nuts (63.875)

NOTE

Connecting rod bearings are available in four undersizes: 0.01 (0.25 mm), 0.02 (0.5 mm), 0.03 (0.75 mm) and 0.04 (1.00 mm).

Component	Minimum	Maximum	Max Wear
Bearing bore (bearing installed and proper torque)	2.3637 (60.040)	2.3650 rod nuts tightened to the (60.071)	Refer to bearing to crank-shaft clearance
Bearing to crankshaft clearance	0.0012 (0.030)	0.0032 (0.081)	0.006 (0.15)
Piston pin bushing bore (less ing)	1.306 (33.19)	1.307 bush- (33.22)	1.3078 (33.22)
Piston pin bushing bore (bushing installed)	1.1820 (30.025)	1.1824 (30.035)	Refer to connecting rod bushing clearance

CAM-

Component	Minimum	Maximum	Max Wear
Bearing journal diameter - front	1.8110 (46.000)	1.8115 (46.014) bearing clearance	Refer to camshaft
Bearing journal diameter - rear	0.9842 (25.000)	0.9848 (25.014)	Refer to camshaft bearing clearance
Camshaft bearing clearance - front	0.0013 (0.035)	0.0039 (0.099)	0.006 (0.15)

Table 1-1. Repair and Replacement Standards - continued.

| Component | Manufacturer Dimensions and Tolerances | | Maximum Allowable |
| | Minimum | Maximum | |
	Inches (mm) Wear	Limit	
Camshaft bearing clearance - rear	0.0010 (0.026)	0.0035 (0.090)	0.006 (0.15)
End play	0.0027 (0.07)	0.041 (1.05)	0.041 (1.05)
Intake lobe- base to tip	1.481 (37.63)	1.491 (37.87)	1.476 (37.5)
Exhaust lobe - base to tip	1.447 (36.76)	1.457 (37.00)	1.443 (36.65)
Injection lobe - base to tip	1.493 (37.91)	1.539 (38.09)	1.488 (37.8)
Transfer pump cam diameter	1.377 (34.98)	1.382 (35.11)	1.358 (34.5)
Transfer pump push rod length	1.856 (47.15)	1.860 (47.25)	1.840 (46.75)

CYLINDER BARREL

Component	Minimum	Maximum	Maximum Allowable
Cylinder bore honed diameter	3.740 (95.000)	3.741 (95.025) (95.075)	3.743
Cylinder bore out of round			0.002 (0.05)
Cylinder bore taper			0.003 (0.075)

PISTON

NOTE

Oversize pistons available +0.0197 in. (0.500 mm)

Component	Minimum	Maximum	Maximum Allowable
Clearance in cylinder (measured from piston pin,1.752 in. (44.5 mm) above bottom of skirt)	0.0102 (0.261)	0.0119 (0.304)	0.0150 90° (0.380)

Table 1-1. Repair and Replacement Standards - continued.

| Component | Manufacturer Dimensions and Tolerances | | Maximum Allowable Wear Limit |
| | Minimum | Maximum | |
	Inches (mm)		
Piston pin bore	1.1812 (30.003)	1.1814 (30.009) piston pin	Refer to clearance in piston
Width of ring groove (top pression)	0.1212 (3.08)	0.1220 (3.10)	0.1240 com- (3.15)
Width of ring groove (2nd pression)	0.100 (2.54)	0.1008 (2.56)	0.1027 com- (2.61)
Width of ring groove (oil control)	0.1590 (4.04)	0.1598 (4.06) (4.11)	0.1618

PISTON PIN

Component	Minimum	Maximum	Maximum Allowable Wear Limit
Diameter	1.1809 (29.995)	1.1811 (30.000) rod bushing	Refer to connecting clearance
Clearance in piston	0.0001 (0.003)	0.0005 0.001 (0.014) (0.025)	
Clearance, connecting rod bushing	0.0009 (0.025)	0.0015 (0.040) (0.050)	0.0019

PISTON RING

NOTE

Oversize piston rings available
+0.0197 in. (0.500 mm)

Component	Minimum	Maximum	Maximum Allowable Wear Limit
Gap (top compression)	0.015 (0.40)	0.025 (0.65) (1.00)	0.039
Gap (2nd compression)	0.015 (0.40)	0.025 (0.65) (1.00)	0.039
Gap (oil control)	0.011 (0.30)	0.023 (0.60) (1.00)	0.039

Table 1-1. Repair and Replacement Standards - continued.

	Manufacturer Dimensions and Tolerances		
	Minimum	**Maximum**	**Maximum Allowable**

VALVE, INTAKE			
Side clearance, top ring	0.0035 (0.09)	0.0055 (0.14)	0.0078 (0.20)
Side clearance, 2nd ring	0.0019 (0.05)	0.0039 (0.10)	0.0059 (0.15)
Side clearance, oil ring	0.0019 (0.05)	0.0039 (0.10)	0.0059 (0.15)
Stem diameter (plated)	0.352 (8.95)	0.353 (8.97)	Refer to stem to guide clearance
Stem to guide clearance	0.0012 (0.030)	0.0029 (0.074)	With valve open 0.29 in. (7.6 mm), side to side movement of valve head not to exceed 0.01 in. (0.25 mm)
Valve seat angle	30°		
Valve face angle	29° 30' ± 15'		
Top of valve head extension above and retraction below cylinder head deck	+0.009 +(0.23)	-0.009 -(0.23)	-0.048 -(1.23)
Valve seat to guide run out		0.002	Regrind to within
VALVE, EXHAUST			
Stem diameter (plated)	0.351 (8.93)	0.352 (8.95)	Refer to stem to guide clearance
Stem to guide clearance	0.002 (0.050)	0.0037 (0.094)	With valve open 0.29 in. (7.6 mm), side to side movement of valve head not to exceed 0.014 in. (0.35 mm)

Table 1-1. Repair and Replacement Standards - continued.

Component	Manufacturer Dimensions and Tolerances Minimum Maximum Inches (mm)		Maximum Allowable Wear Limit
Valve seat angle	45°		
Valve face angle 44° 30' ± 15'			
Top of valve head extension above retraction below cylinder head	+0.009 +(0.23)	-0.009 -(0.23)	-0.048 and -(1.23) deck
Valve seat to guide run-out		0.002 (0.05)	Regrind to within specification

VALVE GUIDES -INTAKE AND EXHAUST

Component	Minimum Inches (mm)	Maximum Inches (mm)	Maximum Allowable Wear Limit
Length	2.24 (57.1)	2.28 (57.9)	
Outside diameter	0.5535 (14.060)	0.5540 (14.074)	
Inside diameter (after reaming)	0.354 (9.000)	0.0355 (9.024)	Refer to stem to guide clearance
Height above valve spring counter bore	0.697 (17.7)	0.705 (17.9)	

VALVE LIFTER

Component	Minimum Inches (mm)	Maximum Inches (mm)	Maximum Allowable Wear Limit
Body diameter	0.9832 (24.973)	0.9837 (24.986)	Refer to valve lifter clearance in bore
Overall length	1.838 (46.7)	1.862 (47.3)	
Clearance in bore	0.0023 (0.058)	0.0040 (0.101)	0.006 (0.150)

Table 1-1. Repair and Replacement Standards - continued.

	Manufacturer Dimensions and Tolerances		Maximum Allowable
	Minimum	Maximum	

VALVE SPRINGS- INTAKE AND EXHAUST

Free length (approximately)	2.25 (57.2)		
Valve spring length - valve open	1.366 (34.7)		
Valve spring length - valve closed	1.732 (44.0)		
Spring load @ 1.732 in. (44.0 mm) - closed	73.1 lb (325 N)	78.9 lb (351 N)	65.8 lb valve (292.5 N)
Spring load @ 1.366 in. (34.7 mm) - valve open	123.7 lb (550 N)	135.3 lb (602 N)	112.4 lb (500 N)

OIL PUMP BODY

Shaft bore diameter (drive)	0.4744 (12.050)	0.4750 (12.065)	Refer to drive shaft clearance in body
Shaft bore diameter (idler)	0.4687 (11.905)	0.4693 (11.920)	
Pump gear bore diameter	1.697 (43.11)	1.699 (43.15)	Refer to oil pump gear radial clearance in body bore
Pump gear bore depth	0.450 (11.44)	0.452 (11.49)	Refer to oil pump gear end clearance to body

1-14

MARINE CORPS TM 05926B/06509B-34
ARMY   TM 5-6115-615-34
NAVY   NAVFAC P-8-646-34

Table 1-1. Repair and Replacement Standards - continued.

| Component | Manufacturer Dimensions and Tolerances | | Maximum Allowable Wear Limit |
	Minimum Inches (mm)	Maximum	
OIL PUMP SHAFTS			
Length (drive)	2.165 (55.0)	2.205 (56.0)	
Diameter (drive)	0.4732 (12.020)	0.4734 (12.025)	Refer to drive shaft clearance in body
Length (idler)	1.634 (41.5)	1.673 (42.5)	
Diameter (idler)	0.4707 (11.955)	0.4713 (11.970)	Refer to idler shaft clearance in gear
Drive shaft clearance in body	0.0010 (0.025)	0.0018 (0.045) (0.076)	0.0030
Idler shaft clearance in gear	0.0006 (0.015)	0.0018 (0.045) (0.060)	0.0024
OIL PUMP GEARS			
Outside diameter (both)	1.6919 (42.975)	1.6929 (43.000)	
Length (both)	0.445 (11.30)	0.446 (11.33)	Refer to end clearance to body
Radial clearance in body bore	0.0043 (0.110)	0.0069 (0.175) (0.220)	0.0087
Inside diameter (both)	0.4719 (11.985)	0.4724 (12.00)	Refer to idler shaft clearance in gear
Gear end clearance to body	0.0043 (0.110)	0.0075 (0.190) (0.220)	0.0087

Table 1-1. Repair and Replacement Standards - continued.

| Component | Manufacturer Dimensions and Tolerances | | Maximum Allowable Wear Limit |
	Minimum Inches (mm)	Maximum	
GEAR BACK-			
Crankshaft to camshaft	0.0024 (0.061)	0.010 (0.267) (0.396)	0.0156
Oil pump to crankshaft	0.0024 (0.061)	0.0131 (0.333)	0.0187 (0.475)
Oil pump element gears	0.0067 (0.170)	0.0168 (0.428)	0.0188 (0.480)
Governor drive to camshaft	0.0024 (0.061)	0.026 (0.661)	0.031 (0.795)
FUEL INJECTOR			
Injector nozzle opening pressure	3117.5 psi 3262.5 psi Re-shim to (215 bar) (225 bar) specifications		

i.   Torque Values.   Refer to Table 1-2. for critical torque values and all engine torque values. All other items are tightened to standard torques, refer to Table 1-3.

Table 1-2.Critical Torque Values.

Location	Fastener Size/Class	lb ft	(N.m)
Flywheel	M10x1.5/12.9	57-58	(77-79)
Pushrod shield adapter	M6x1.0/8.8	7-8	(10-11)
Oil pan	M6x1.0/8.8	7-8	(10-11)
Oil pan drain plug	M18X1.5	31-32	(41-43)
Rocker arm studs ###	M10x1.5/10.9	28-30	(38-40)
Injector clamp stud ###	M8x1.25/10.9	5-6	(7-8)
Injector clamp nut ##	M8x1.25/10.9	7-8	(9-11)
Cylinder head nuts ##, ***	MJ10x1.5/10.9	33	(45) Cylin-
der head studs ###	MJ10x1.5/12.9	Refer to installation procedures	

MARINE CORPS TM 05926B/06509B-34
ARMY   TM 5-6115-615-34
NAVY   NAVFAC P-8-646-34
AIR FORCE TO 35C2-3-386-32

Table 1-2.Critical Torque Values - continued.

Location	Fastener Size/Class	lb ft	(N.m)
Valve cover	M8x1.25/8.8	12-13	(16-18)
Breather cover	M6x1.0/10.9	7-8	(10-11)
Mounting feet	M8x1.25/8.8	16-18	(21-24)
Intake manifold	M8x1.25/8.8	16-18	(21-24)
Exhaust manifold stud ##, *	M8x1.25/9.8	5-6	(7-8)
Exhaust manifold nuts ##	M8x1.25/8.8	16-18	(21-24)
Sound shield & scroll mounting ##	M6x1.0/8.8	4-5	(6-7)
Sheet metal	M6x1.0/8.8	5-6	(7-8)
Lifting eye nut	M8x1.25/8.8	16-18	(21-24)
Lifting eye stud	M8x1.25/9.8	5-6	(7-8)
Fuel injection pump	M8x1.25/10.9	24-26	(32-35)
Fuel transfer pump #	M6x1.0/8.8	7-8	(10-11)
Injection line fittings #	M12x1.5	16-18	(21-24)
Fuel line banjo bolts ##	M12x1.5	18-20	(24-27)
Generator adapter	M10x1.5/8.8	30-32	(41-43)
Connecting rod nuts ##, **	M10x12.5/12.9	60-66	(81-89)
Oil pump cover plate	M6x1.0/8.8	6-7 7-8	(8-9)
Oil pump mounting	M6x1.0/8.8	7-8	(10-11)
Camshaft retainer	M6x1.0/8.8	7-8	(10-11)
Gear cover	M6x1.0/8.8	1-2	(10-11)
Alternator	M4x0.7/8.8	2-3	(2-3)
Governor drive gear	M4x0.7/12.9	7-8 5-6	(3-4)
Governor	M6x1.0/12.9	7-8	(10-11)
Fan mounting	M6x1.0/8.8	12-13	(7-8)
Flexplate mounting	M6x1.0/8.8	60-62	(10-11)
Glow plug	M10x1.0	7-8	(16-18)
Starter to starter mount	M12x1.75/10.9	16-18	(82-84)
Starter to support bracket	M6x1.0/8.8	20-25	(10-11)
Starter mount to crankcse	M8x1.25/8.8	20-25	(21-24)
Oil pump bypass plug	M14x1.5	4-5	(27-34)
Oil filter adapter ###	M16x1.5	7-8	(27-34)
Injection pump actuator shaft	M6X1.0.	7-8	(6-7)
Solenoid mounting bracket	M6x1.0/8.8	7-8	(10-11)
Solenoid mounting	M6x1.0/8.8		(10-11)
Governor cable mounting clip	M6x1.0/8.8		(10-11)

NOTES: Use sealing compound (Table 2-2, item 7) on all fasteners except as indicated below. Recommended torque. Torquing not required.

\#   Do not use sealing compound.Use may be detrimental.

\##   Use sealing compound (Table 2-2,item 5).

\###   Use sealing compound (Table 2-2, item 6) on cylinder head end only.

*

* Before torquing, apply lube oil to
  bearing faces on nut and rod cap.

* Before torquing, apply lube oil to and finally 45 N.m. (15, 26, 33 ft 3-4. See Figure 1-1.
bolt and nut threads and to adjacent load

stud and nut threads.Torque nuts to 20, 35, lb) using the sequence 1-2-3-4-4-3-2-1-1-2-

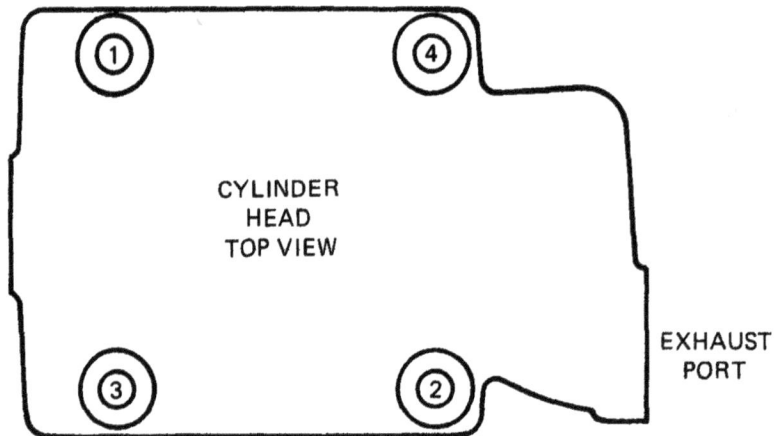

Figure 1-1.Cylinder Head Torquing Sequence.

Table 1-3.Standard Torque Values.

Approximate Screw Size	Torque in.
1/4 (6)	3-5 (4-7)
5/16 (8)	7-11 (10-15)
3/8 (10)	14-18 (19-24)
7/16 (11)	23-28 (31-38)
1/2 (13)	32-37 (44-50)

1-13. DIFFERENCES BETWEEN MODELS. This manual covers DOD Models MEP-016B, MEP-021B, and MEP-026B.The differences that exist between these models are discussed in the appropriate areas in this manual. The basic difference between models is MEP-016B delivers 60 hertz Alternating Current (AC), MEP-021B delivers 400 hertz (AC), and the MEP-026B delivers 28 volts Direct Current (DC).

CHAPTER 2

GENERAL MAINTENANCE INSTRUCTIONS

Section I. REPAIR PARTS, SPECIAL TOOLS, TEST, MEASUREMENT, AND DIAGNOSTIC EQUIPMENT (TMDE), AND SUPPORT EQUIPMENT

2-1. REPAIR PARTS. Repair parts are listed and illustrated in TM 05926B/06509B-24P/ TM 5-6115-615-24P/TO 35C2-3-386-24P/NAVFAC P-8-646-24P parts manual for the 3kw Generator Set.

2-2. TOOLS AND EQUIPMENT. Table 2-1 contains a list of all special tools, test and support equipment needed to maintain this unit.

Table 2-1. Special Tools, Test and Support Equipment.

Item	NSN Reference Number Reference (Or Equivalent)	Para. No.	Use	
Valve Spring Compressor		7-7 Tim-7-10 7-10	Remove/Install Engine Timing	Valves
High Pressure Pump scilloscope	bearing 72-7010 6-3		Injection Pump Waveform Analysis	Timing
Hydraulic Press (with seal and installation adapters)		7-12,7-17,7-19	Seal and Bearing Installation	
High Potential Insulation Test (Megohmmeter)		5-4,5-6,5-7	Generator Windings	
Valve Grinder		7-7	Valve Repair	
Post Mold (+)			Battery Repair	
Post Mold (-)	512-00-251-5045 4-2 5120-00-251-5046 4-2		Battery Repair	
Power Supply (0-150VAC)		8-2, 5-9	Voltage Regulator Test	

MARINE CORPS TM 05926B/06509B-34
ARMY   TM 5-6115-615-34
NAVY   NAVFAC P-8-646-34
AIR FORCE   TO 35C2-3-386-32

Table 2-1. Special Tools, Test and Support Equipment - Continued

Item	(Or Equivalent) NSN Reference Number	Reference Para. No.	Use
Power Supply (Variable DC)		9-3	Frequency Meter
Epoxy Kit	8010-00-959-4661	4-2	Battery Repair
Multimeters (3)		5-2, 5-4, 5-5 5-6, 5-7, and 8-2	Circuit Testing
Signal Generator		8-2	Circuit Testing
Rheostat 750 ohm, 2 Watt		5-9	Voltage Regulator Testing/Adjustment
Resistor 30 ohm, 120 Watt		5-9	Voltage Regulator Testing/Adjustment
Potentiometer 500 ohm		5-9	Voltage Regulator Adjustment
Voltmeter (AC)		5-9	Voltage Regulator Adjustment
Voltmeter (DC)		5-9	Voltage Regulator Adjustment
Cylinder Hone		7-13	Remove Cylinder Glaze
Nozzle Cleaning Kit	KDEP-1043	6-4	Injection Spray Hole Cleaning
Nozzle Tester	0-681-143-014	6-4	Injector Testing, Setting Nozzle Opening Pressure
Rocker Arm Socket	420-0517	7-8	Remove/Install Rocker Arm Nuts
Ring Expander		7-13	Remove/Install Rings

Crankshaft Gear:
  Puller
  Socket
	PS87-104-1
	PS87-104-2-2
		7-17	Remove Crank Gear

MARINE CORPS TM 05926B/06509B-34
ARMY   TM 5-6115-615-34
NAVY   NAVFAC P-8-646-34
AIR FORCE TO 35C2-3-386-32

Table 2-1.Special Tools, Test and Support Equipment - Continued

Item	NSN Reference Number (Or Equivalent)	Reference Para. No.	Use
Load Bank		7-2	Governor Adjustment
Bar Stock	420-0518	7-11	Setting Timing Pointer

2-3.FABRICATED TOOLS for **AND EQUIPMENT.** No fabricated tools or equipment are necessary 3kw Generator Sets. the .

Table 2-2. Consumable Operating and Maintenance Supplies.

Item No.	Component Application	National Stock Number	Required Description	Required	Qty For Initial Operation	Qty 8 Hours Operation	Notes
1	Battery	8010-00-959-4661	Epoxy Kit		As req'd		
2	Battery	9650-00-264-5050	Pig Lead		As req'd		
3	Starter		Grease Multitemp PS No. 2	As req'd			
4	Lifter		Lifter, Pre-lube Type, sealed power LL-S	As req'd			
5	Misc.	8030-00-148-9833	Loctite 271 MIL-S-46163-A Type I, Grade K		As req'd		
6	Misc.		Never-Seeze (Onan P/N 524-0076)		As req'd		
7	Misc.		Loctite 242 MIL-S-46163A Type II, Grade N		As req'd		
8	Misc.		Sealing Compound MIL-S-46163-A Type III, Grade R Removable	As req'd			
9	Misc.		Sealing Compound, RTV Silicone MIL-A-46106-A	As req'd			
10	Misc.		Loctite Ultra Blue Silicone Sealant		As req'd		
11	Misc.		Sealing Compound MIL-R-46082 Type I	As req'd			
12	Misc.		Dry Cleaning Solvent, P-D-680		As req'd		

Item Number	Component Description	National Stock Operation	Required For Initial No.	Qty Application	Qty Required 8 Hours Operation	Motes
13	Misc. Sealing compound As req'd		MIL-S-22473-E Grade HVV			
14	Misc.	P/N 13217E3704, FSCM 97403	Heat Sink Compound	As req'd		

Section II.TROUBLESHOOTING

2-4.GENERAL. This section contains troubleshooting information for locating and correcting operating troubles which may develop in the Generator Set.Each

malfunction for an individual component, unit or system is followed by a list of tests or inspections which will help you determine probable causes and corrective actions to take. You should perform the tests/inspections and corrective actions in the order listed.

2-5.MALFUNCTIONS NOT CORRECTEDBY USE OF THE TROUBLESHOOTING TABLE. This manual cannot list all malfunctions that may occur, nor all tests or inspections and corrective actions. If a malfunction is not listed or cannot be corrected by the listed corrective actions, notify your supervisor.

Table 2-3. Troubleshooting.

Malfunction
  Test or Inspection
    Corrective Action

---

1. STARTER MOTOR DOES NOT TURN - HIGH CURRENT DRAW.

Step 1. Check   starter for grounded solenoid. to paragraph 7-4.c.
      Refer   and 7-5.a.

              Replace any grounded component.

Step 2. Check   starter for grounded terminals. to paragraph 7-4.c.
      Refer

              Replace any grounded component.

              starter for grounded field stator. to paragraph 7-4.c.
Step 3. Check
      Refer   Replace any grounded component.

              for frozen armature shaft.Refer to paragraph 7-4.d.e.

Step 4. Check   Replace armature bushing and, if necessary, armature.

2. STARTER DOES NOT TURN - NO CURRENT DRAW.

Step 1. Check for open   armature windings.Refer to paragraph 7-4.b.c.

              Replace   armature if necessary.

Step 2. Check for open   starter field windings. Refer to paragraph 7-4.c.

              Replace   field windings if necessary.

Table 2-3.Troubleshooting - Continued.

Malfunction		
Test or Inspection		
	Corrective Action	

Step 3. Check   for broken or weak brush springs. Refer to paragraph 7-4.d.

    Replace brush springs if necessary.

Step 4. Check   for worn commutator (high mica).   Refer to paragraph 7-4.d.

    Refinish commutator or replace armature (including commutator).

3. SLOW STARTER SPEED.

Step 1.   Check   for dirty commutator.   Refer to paragraph 7-4.d.

    Clean or refinish commutator.

Step 2.   Check   for worn armature shaft bushings.   Refer to paragraph 7-4.d.

    Replace bushings and, if necessary, armature.

Step 3.   Check   for burned solenoid contacts.   Refer to paragraph 7-4.d.

    Replace solenoid.

Step 4.   Check   for open or shorted starter field windings.   Refer to paragraph 7-4.c.

    Replace windings, if necessary.

Check for worn generator end bearing.   Refer to paragraph 5-8.a.

Step 5.   Replace bearing if necessary.

4. STARTER WILL NOT ENGAGE FLYWHEEL.

Step 1. Check   starter free running clutch. Refer to paragraph 7-4.d.

    Replace clutch if necessary.

Step 2. Check   starter pinion gear.Refer to paragraph 7-4.d.

    Replace pinion if necessary.

Step 3. Check   for damaged flywheel ring gear. Refer to paragraph 7-1.b.

    Replace flywheel if necessary.

Table 2-3. Troubleshooting - Continued.

Malfunction

    Test or Inspection

        Corrective Action

5. ENGINE WILL NOT START WHEN CRANKED.

Step 1.Check glow plug operation.Refer to TM 05926B/06509B-12/TM 5-6115-615-12/NAVFAC P-8-646-12/TO 35C2-3-386-31 manual, paragraph 4-50.a.

Replace a defective glow plug or wiring.

Step 2. Check for faulty injection caused by dirty fuel.   Refer to paragraph 6-4.c.

Replace with clean fuel.   Service fuel filters, refer to TM 05926B/06509B-12/ TM 5-6115-615-12/NAVFAC P-8-646-12/TO 35C2-3-386-31 manual, paragraph 4-50. Clean and, if necessary, repair or replace fuel injection nozzle.Refer to paragraph 6-4.e.f.g.

Step 3. Check for poor compression.   Refer to paragraph 7-2.a.

See malfunction 12.

Step 4. Incorrect fuel injection pump   timing.   Refer to paragraph 6-2.e.

Retime pump.Refer to   paragraph 6-2.e.

6. ENGINE MISFIRES.

Step 1.Check for poor compression.   Refer to paragraph 7-2.a.

See malfunction 12.

Step 2. Check for defective or dirty injection   nozzle.   Refer to paragraph 6-4.c.

Clean or replace nozzle. Refer to paragraph 6-4.e.f.g.

Step 3. Check for broken valve springs.Refer   to paragraph 7-7.a.

Replace broken springs.Refer to paragraph 7-7.c.

Step 4. Check cylinder head assembly for build-up of carbon.   Refer to paragraph 7-6.c.

Clean carbon from cylinder head assembly. Refer to paragraph 7-6.c.

Malfunction
    Test or Inspection
        Corrective Action

---

7. LOW ENGINE POWER.

Step 1. Check   for poor compression.Refer to paragraph 7-2.a.

    See malfunction 12.

Step 2. Check   for dirty or defective injection nozzle.   Refer to paragraph
    6-4.c

    Clean or replace injection nozzle.   Refer to paragraph 6-4.e.f.g.

Step 3. Check   for incorrect fuel injection pump timing. Refer to paragraph
    6-2.e.   injection timing if necessary.   Refer to paragraph

    Adjust
    6-2.e.

8. EXCESSIVE OIL CONSUMPTION.

Step 1. Check   for worn valve guides.   Refer to paragraph 7-6.c.

    Replace guides, valves and seals.Refer to paragraph 7-6.d for guide replacement and paragraph 7-7.b for valve and seal replacement.

Step 2. Check   for worn or sticking piston rings.   Refer to paragraph 7-
13.b.

    Replace rings if necessary.Refer to paragraph 7-13.c.d.

9. BLACK SMOKEY EXHAUST AND EXCESSIVE FUEL CONSUMPTION

Step 1. Check   for incorrect fuel injection pump timing. Refer to paragraph
    6-2.e.

    Adjust injection timing if necessary. Refer to paragraph 6-2.e.

Step 2. Check for faulty injection pump. Refer to paragraph 6-2.e.

    Replace pump as necessary.

Step 3. Check for faulty fuel injection nozzle.   Refer to paragraph 6-4.c.

    Repair or replace nozzle.See paragraph 6-4.e.f.g.

Table 2-3.Troubleshooting - Continued.

Malfunction

    Test or Inspection

        Corrective Action

Step 4.Check valve condition.Refer to paragraph 7-7.a.

    Repair or replace worn valves.   Refer to paragraph 7-7.b.c.

10. TAPPING OR CLICKING SOUND FROM CYLINDER HEAD.

Step 1.Check for excessive valve clearance.   Refer to paragraph 7-8.e.

    Adjust clearance.

Step 2. Check for broken valve spring. Refer   to paragraph 7-7.a.

    Replace broken springs. Refer   to paragraph 7-7.c.

11. METALLIC KNOCKING, CLICKING, OR POUNDING FROM CRANKCASE OR CYLINDER BLOCK.

Step 1.   Check for dirty or defective injection nozzle. Refer to paragraph 6-4.c.

    Clean or replace injection nozzle.   Refer to paragraph 6-4.e.f.g.

Check for worn or loose connecting rod bearings.Refer to paragraph 7-15.c.

Step 2.   Replace bearings if necessary.

Check for loose piston assembly.Refer to paragraph 7-13.c.

    Repair or replace piston.Refer to paragraph 7-13.d.e.

Step 3.

Check for loose connecting rod assembly. Refer to paragraph 7-15.C.

    Replace connecting rod bearings.

Step 4.

12. LOW ENGINE COMPRESSION.

NOTE

Normal cylinder pressure is between 325 and 375 psi depending upon engine condition. Maintenance should be considered if pressure is below 325 psi.

MARINE CORPS TM 05926B/06509B-34
ARMY   TM 5-6115-615-34
NAVY   NAVFAC P-8-646-34
AIR FORCE   TO 35C2-3-386-32

Table 2-3. Troubleshooting - Continued.

Malfunction

   Test or Inspection

      Corrective Action

   Step 1. Check   for loose cylinder head.

              Properly tighten head.   Refer to paragraph 7-6.f.

   Step 2. Check   for broken valve spring.   Refer to paragraph 7-7.a.

              Replace broken spring.   Refer to paragraph 7-7.c.

   Step 3. Check   for worn or sticking valves. Refer to paragraph 7-7.b.

              See malfunction 13 below.

   Step 4. Check   for worn valve seats. Refer to paragraph 7-6.c.

              Clean and, if necessary, regrind valve seats.   Refer to
              paragraph 7-6.d.

   Step 5. Check for worn or sticking piston rings.Refer to paragraph
        7-13.C.

              Replace rings if necessary.Check condition of cylinder walls and piston grooves.

   Step 6. Check for worn cylinder walls and pistons.Refer to paragraph
        7-13.C.

              Refinish cylinder walls. Replace pistons.

1. STICKING VALVES.

   Step 1.Check for dirty, scored, or gummy valve stems or guides.   Refer to
        paragraph 7-6.c.

              Clean stems and guides. Replace guides if necessary.
              Refer to paragraph 7-6.c.d.

   Step 2. Check for weak or broken springs. Refer to paragraph 7-7.a.

              Replace springs.   Refer to paragraph 7-7.c.

1. ENGINE RUNS NORMALLY, BUT GENERATOR HAS NO OUTPUT.

   Step 1.Check voltage regulator.   Refer to paragraph 5-9.a.

              Repair or replace regulator.Refer to paragraph 5-9.d.e.

Table 2-3. Troubleshooting - Continued.

Malfunction

    Test or Inspection

        Corrective Action

        Step 2. Check exciter field (stator) for open or shorted windings.   Refer
           to paragraph 5-7.

               Replace stator if necessary. Refer to paragraph 5-2.e.f.
    Check diodes on exciter rotor. Refer to paragraph 5-5.a.

        Step 3. Check   Replace diodes if necessary. Refer to paragraph 5-5.b.c.

               generator field (rotor) for open or shorted windings. Refer
               to paragraph 5-4.b.

        Step 4. Check   Replace rotor if necessary. Refer to paragraph 5-4.c.
       generator stator for open, shorted, or grounded windings to paragraph 5-6.a.

            Replace stator, if necessary. Refer to Paragraph 5-2.f.

        Step 5. Check   exciter rotor for open, shorted, or gorunded windings. Refer
           Refer

        Step 6. Check
           to paragraph 5-4.b.

             Replace exciter rotor if necessary. Refer to paragraph 5-2.f.

        Step 7. Check bridge assembly (MEP-026B Only) for defective diodes.Refer to TM 05926B/06509B-12/TM 5-6115-615-12/NAVFAC P-8-646-12/TO 35C2-3-386-31 manual, paragraph 69.c.

             Replace defective diodes.Refer to TM 05926B/06509B-12/TM 5-6115-615-12/NAVFAC P-8-646-012/TO 35C2-3-386-31 manual, paragraph 4-69.d.

15. ENGINE RUNS NORMALLY, BUT GENERATOR HAS LOW OUTPUT.

        Step 1.Check voltage regulator.   Refer to paragraph 5-9.a.

             Adjust, repair, or replace regulator. Refer to paragraph 5-9.d.e.

        Step 2.   Check voltage adjust rheostat.   Refer to TM 05926B/06509B-12/TM 5-6115-615-1/NAVFAC P-8-646-12/TO 35C2-3-386-31 manual, paragraph 4-65.a.

             Replace rheostat if necessary.   Refer to TM 05926B/06509B-12/TM 5-6115-615-12/NAVFAC P-8-646-12TO 35C2-3-386-31 manual, paragraph 4-65.b.c.

MARINE CORPS TM 05926B/06509B-34
ARMY   TM 5-6115-615-34
NAVY   NAVFAC P-8-646-34
AIR FORCE   TO 35C2-3-386-32

Table 2-3. Troubleshooting - Continued.

Malfunction
    Test or Inspection
        Corrective Action

Step 3. Check generator stator for open or shorted windings. Refer to paragraph 5-6.a.

    Replace stator. Refer to paragraph 5-2.f.

Step 4. Check exciter rotor for shorted windings. Refer to paragraph 5-4.b.

    Replace rotor. Refer to paragraph 5-2.f.

Step 5. Check diodes on exciter rotor. Refer to paragraph 5-5.a.

    Replace diodes as necessary. Refer to paragraph 5-5.c.

16. HIGH GENERATOR OUTPUT VOLTAGE (NO LOAD).

    Step 1. Check for voltage regulator failure. Refer to paragraph 5-9.a.

    Repair or replace defective regulator. Refer to paragraph 5-9.d.e.

    Step 2. Check voltage adjust rheostat. Refer to TM 05926B/06509B 12/TM 5-6115-615-12/NAVFAC P-8-646-12/TO 35C2-3-386-31 manual, paragraph 4-65.a.

    Replace rheostat if necessary. Refer to TM 05926B/06509B-12/TM 5-6115-615-12/NAVFAC P-8-646-12/TO 35C2-3-386-31 manual, paragraph 4-65.b.c.

17. FREQUENCY METER FAILS TO REGISTER.

    Step 1. Test frequency meter. Refer to paragraph 9-3.a.

    Replace defective frequency meter. Refer to TM 05926B/06509B-12/TM 5-6115-615-12/NAVFAC P-8-646-12/TO 35C2-3-386-31 manual, paragraph 4-66.a.b.

Step 2. Test frequency transducer. Refer to paragraph 9-4.a.

    Replace defective frequency transducer. Refer to TM 05926B/06509B-12/TM 5-6115-615-12/NAVFAC P-8-646-12/TO 35C2-3-386-31 manual, paragraph 4-68.b.c.

Table 2-3.Troubleshooting - Continued.

Malfunction
　　　Test or Inspection
　　　　　Corrective Action

18. BATTERIES DO NOT CHARGE.

Step 1.　　Check battery charging alternator stator.　　　　　　　Refer to paragraph 7-11.a.

　　　　　　　Replace defective stator.Refer to paragraph 7-11.b.f.

　　　　　Check flywheel magnets.Refer to paragraph 7-10.b.
Step 2.

　　　　　　　Replace flywheel.Refer to paragraph 7-10.c.

　　　　　Check battery charging voltage regulator.Refer to TM 05926B/06509B-12/TM 5-6115-615-12/NAV-
Step 3.　FAC P-8-646-12/TO 35C2-3-386-31 manual, paragraph 4-25.b.

　　　　　　　Replace defective battery charging voltage regulator. Refer to TM 05926B/06509B-12/TM
　　　　　　　5-6115-615-12/NAVFAC P-8-646-12/TO 35C2-3-386-31 manual, paragraph 4-25.c.d.

Section III.GENERAL MAINTENANCE

2-6.GENERAL MAINTENANCE. This section contains general maintenance instructions which are the responsibility of intermediate direct support and general support maintenance personnel. You will find that these instructions apply to several components or assemblies. chapter.
　　　　　　　They would otherwise have to be repeated throughout the

2-7.　　GENERAL MAITENANCREQUIRE-

2-7.1.Work Guidelines.

　　a. Make sure the work　　　　　area is clean before you disassemble　the generator or
　　　　　engine.

　　b. Make sure that the　　　　　materials needed for the task are at　hand.　　These may
　　　　　include cleaning solvents, lubricants, buckets or other　containers for

cleaning parts or all of the   keeping   components separated, clean wiping cloths, and

---

WARNING

DRY CLEANING
SOLVENT, (Table 2-2, item 12), used to clean parts is potentially dangerous to personnel and property. Avoid repeated and prolonged skin contact. DO NOT use near open flame or excessive heat.Flash point of solvent is 100 to 138 degrees Fahrenheit (38 to 60 degrees Centigrade).

c.   Clean the exterior of the engine and generator before disassembly to keep foreign matter from contaminating bearings, gears and other components which may be damaged by these foreign materials.Use a clean cloth dampened with cleaning solvent (Table 2-2, item 12).

d.   If compressed air is used to clean parts, make sure it is free of dirt and other contaminants.Never exceed 15 psi nozzle pressure.

e.   Protect disassembled parts from dust, blowing sand, and moisture which can

cause rapid wear and deterioration of bearings, parts.   gears and other machined

2-7.2.   Seals and Gaskets.   Replace seals and gaskets of components. This will greatly reduce the possibility of   all disassembled and will help prevent the entry of dust and dirt.   leaking after reassembly,

2-7.3.   Care of Bearings.

a.   Clean ball and roller bearings by placing them in a wire basket and immersing them in a container of fresh cleaning solvent. Agitate the bearings in the solvent to remove all traces of old lubricant.

b. Dip the cleaned bearings in clean engine oil and immediately wrap them in lint-free paper to protect them from dust and other foreign matter.

2-7.4.   **Replacing Electronic Components.**

a. Tag wires with identification before removal to ensure proper reconnection.

b. When soldering, use a heat sink between the soldering pencil and the electronic component to prevent damage to the component.

c. Do not use excessive heat when soldering on printed circuit boards. Damage to the board may result.

2-7.5.   Fuel System Maintenance.   The fuel injectors and fuel pump are manufactured to extremely tight tolerances.Even small amounts of dust or water in the fuel system can damage these components.   It is of great importance that dirt be kept out of the fuel lines and fittings during disassembly and assembly.All openings should be taped, capped or plugged immediately after disassembly.   If dirt does accidentally enter a component, it should be flushed with clean diesel fuel before reassembly.

2-7.6.   <u>Welding.</u>

   a. Aluminum welding.Repair of aluminum components by welding is accomplished in accordance with MIL-W-45205, Class B.

   b. Steel welding. Repair of steel components by welding is accomplished in accordance with MIL-W-8611.

     Resistance welding.Resistance welding is accomplished in accordance with MIL-W-12332.

   c.

     <u>Painting.</u>   Components which require painting are treated and painted in

2-7.7.

accordance with MIL-T-704, Type F or G, NATO Green 383.

Section IV.REMOVAL AND INSTALLATION OF MAJOR COMPONENTS.

2-8.   CONTROL BOX.

  a.<u>Removal.</u>

    (1) Disconnect negative (-) battery cable from battery.

NOTE

      Figure 2-1 shows the Model MEP-016B Control box.
      Models MEP-026B and MEP-021B (28VDC and 400 Hz) are similar.

      Remove screws (1, Figure 2-1) and nuts (2). Remove speed control mounting bracket from bottom of control box.

  (2)

      Remove nut (3), ground strap (4), and screw (5).

    (3) Disconnect five generator leads (MEP-026B) and eight generator leads (MEP-016B and MEP-021B).

    (4)   (a) MEP-026B (28VDC).

         <u>1</u>   Tag and disconnect two exciter stator leads from terminals #1 and #5 of terminal board A3-TB1 (see Figure 2-2).

         <u>2</u>Tag and disconnect three main stator leads T1, T2, and T3.T1 is connected to terminal #4 of terminal board A3-TB1.
          Lead T2 is connected to terminal #2 of A3-TB1 and lead T3 is connected to terminal #3 of A2-TB2.

      (b) MEP-016B and MEP-021B (60 and 400 Hz).

         <u>1</u> Tag and disconnect exciter stator leads F1 and F2 at terminals #1 and #2 of terminal block A1-TB2 (see Figure 2-3).

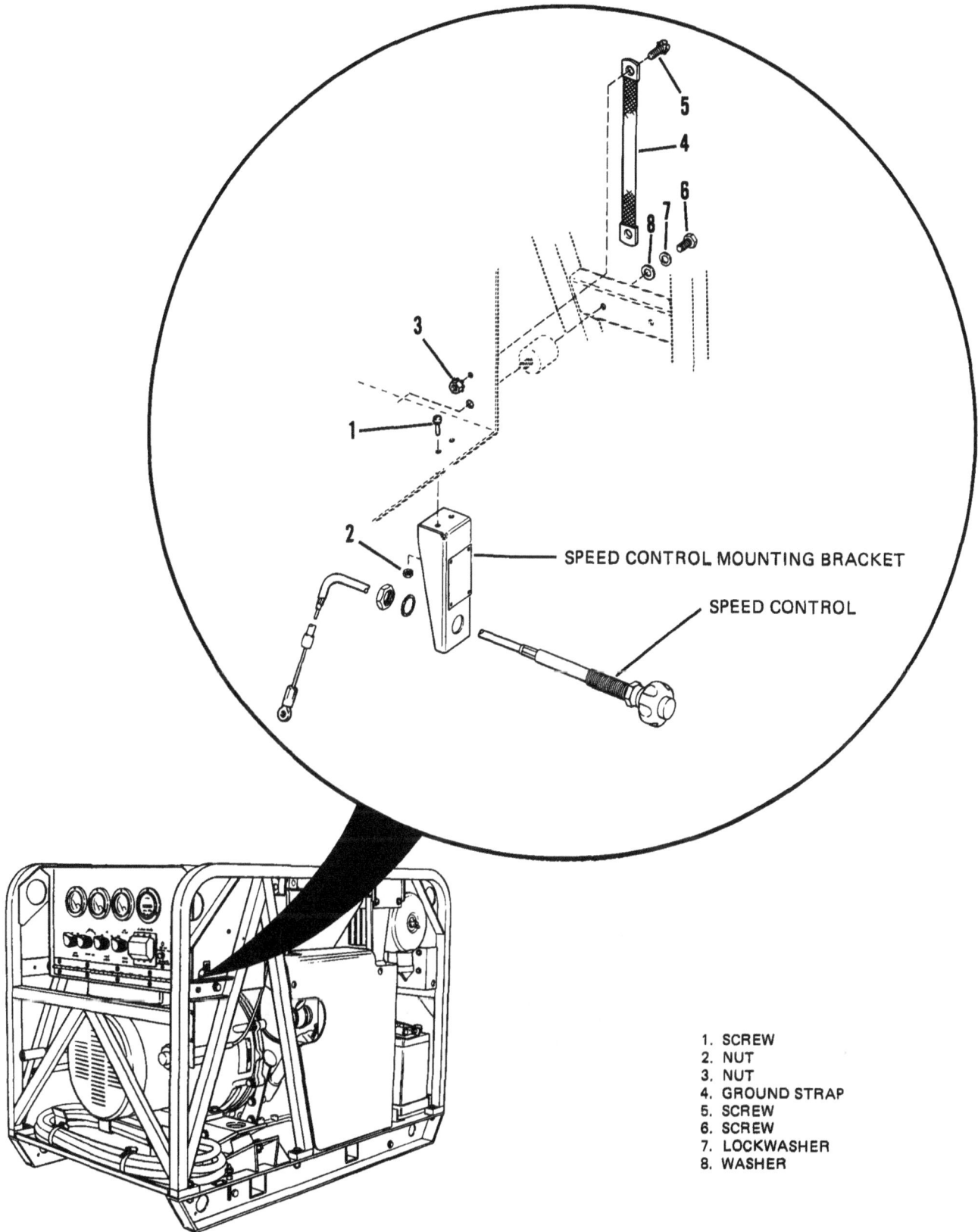

SPEED CONTROL MOUNTING BRACKET

SPEED CONTROL

1. SCREW
2. NUT
3. NUT
4. GROUND STRAP
5. SCREW
6. SCREW
7. LOCKWASHER
8. WASHER

Figure 2-1. Control Box Removal and Installation.

2   Tag and disconnect six main stator leads T1, T2, T3, T4, T5, and T6. Lead T1 is connected to terminal #1 of terminal block A1-TB1, lead T2 is connected to terminal #2 of A1-TB1, lead T3 is connected to terminal #3 of A1-TB1, lead T4 is connected to terminal #4 of A1-TB1, lead T5 is connected to terminal #5 of A1-TB1, and lead T6 is connected to terminal #6 of A1-TB1.

(5) Remove four screws (6, Figure 2-l), lockwashers (7), and washers (8).

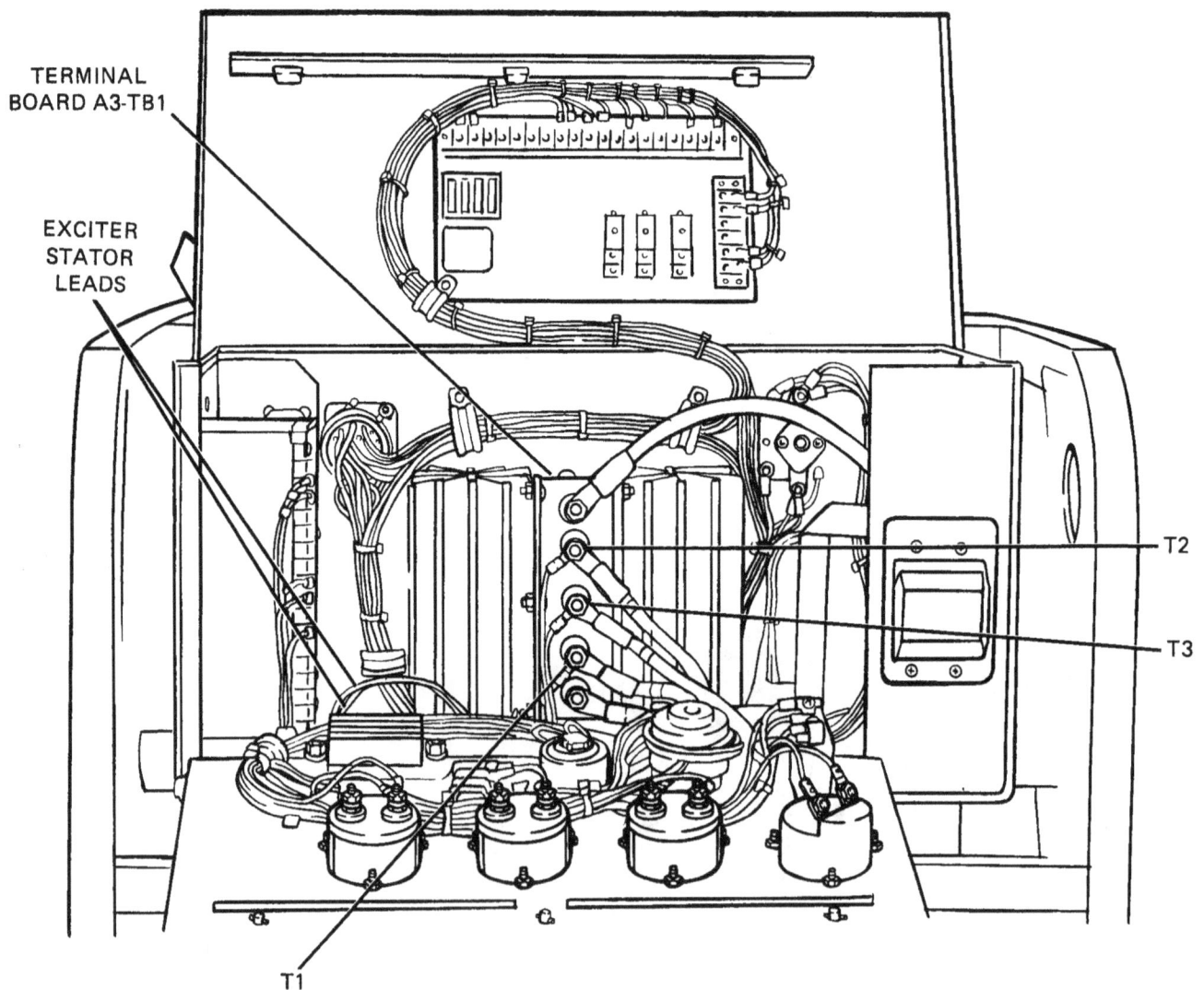

Figure 2-2.Generator Wiring MEP-026B.

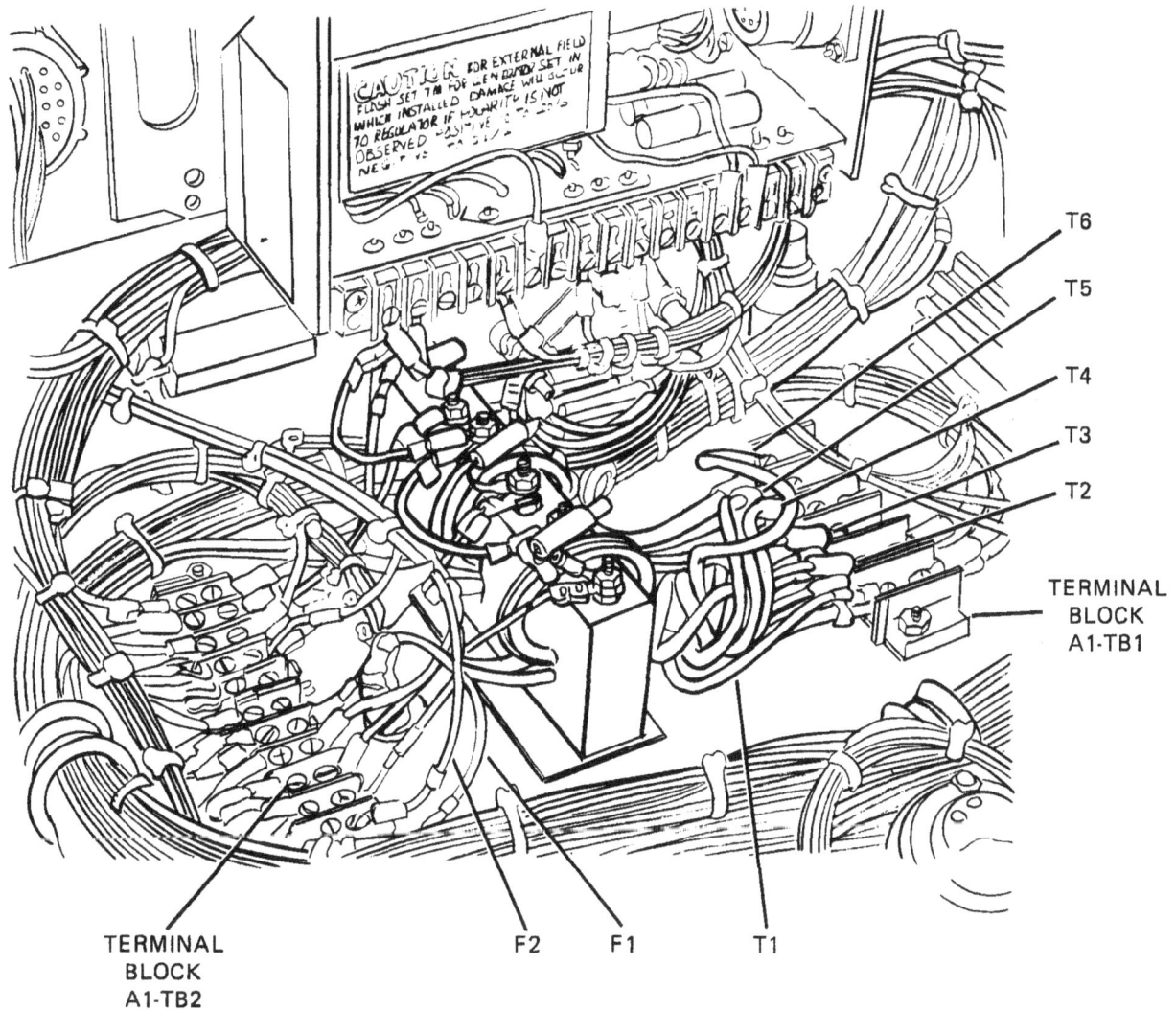

Figure 2-3.Generator Wiring MEP-016B and MEP-021B.

(6) Loosen clamp and disconnect control box harness (Figure 2-4).

(7) Thread generator wiring (five leads MEP-026B,eight leads MEP-016B and MEP-021B) through opening in bottom of control box.

(8) Remove control box and cover generator opening to prevent entry of dirt.

b.Installation.

(1) Remove cover from generator.

(2) Place control box in position on frame.   Thread generator wiring (five leads MEP-026B, eight leads MEP-016B and MEP-021B) through opening in bottom of control box.

(3) Secure control box to frame and screws   with washers (8, Figure2-1), lockwashers (7),
(6).

(4) Connect generator wiring to identification. control box terminal boards using tags as lead

(5) Secure ground strap (4) with screw (5)   and nut (3).

(6) Connect   speed control mounting bracket nuts (2).   to bottom of control box with screws
(1) and

(7) Connect   control box harness to back of 2-4).   control box and tighten clamp
(Figure

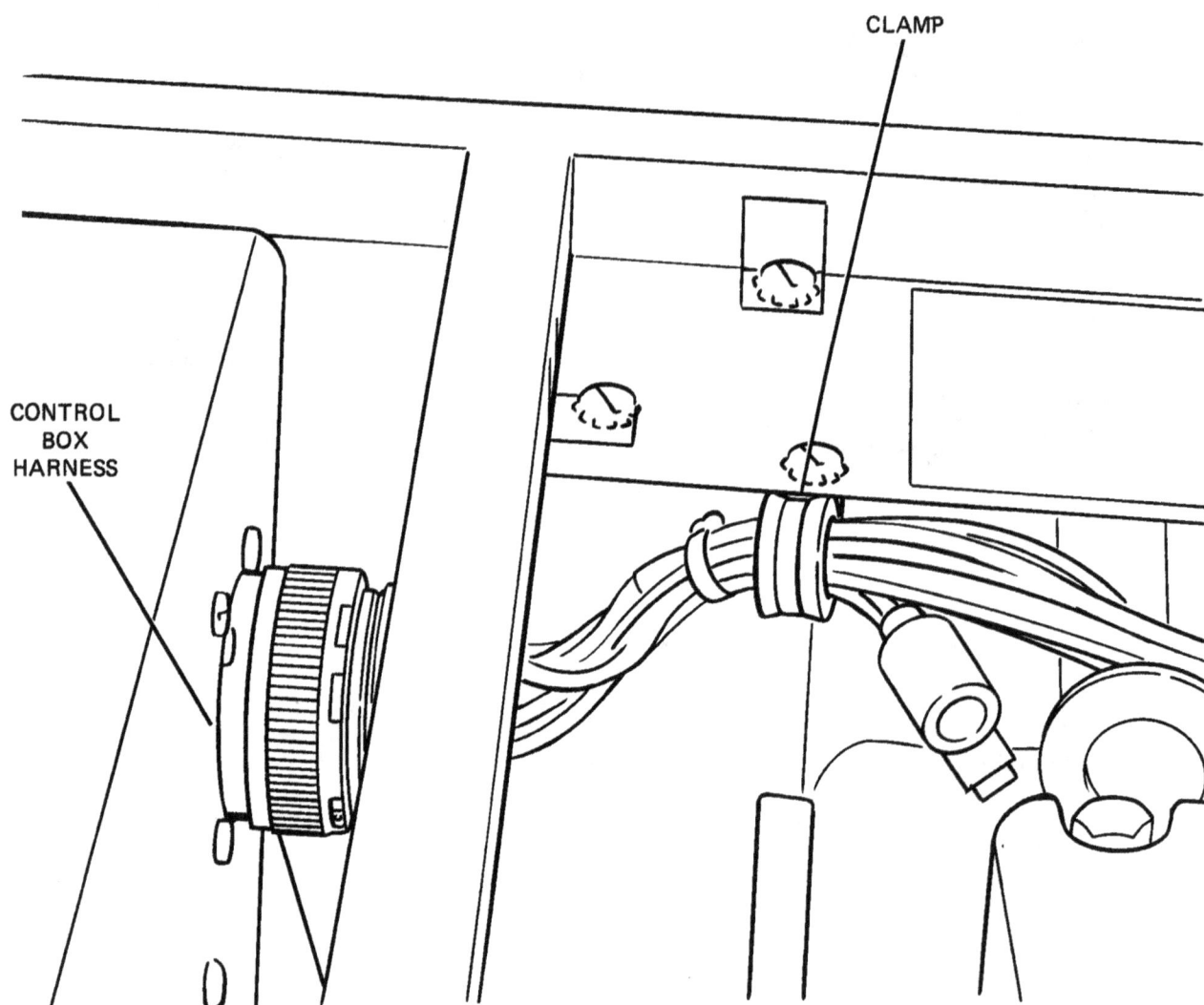

CLAMP

CONTROL
BOX
HARNESS

MARINE CORPS TM 05926B/06509B-34
ARMY   TM 5-6115-615-34
NAVY   NAVFAC P-8-646-34

2-9.   ENGINE.

a. Removal.

(1) Drain fuel from fuel tank.

CAUTION

Do not, apply air pressure tothe engine crankcase to speed the oil drain process. Air pressure can force the oil seals out of the crankcase.

NOTE

To expedite the oil draining process, raise end of set opposite the drain valve.

(2) Open drain valve (1, Figure 2-5), drain the engine oil. Disconnect oil drain line (2) from the drain valve and plug the oil line.

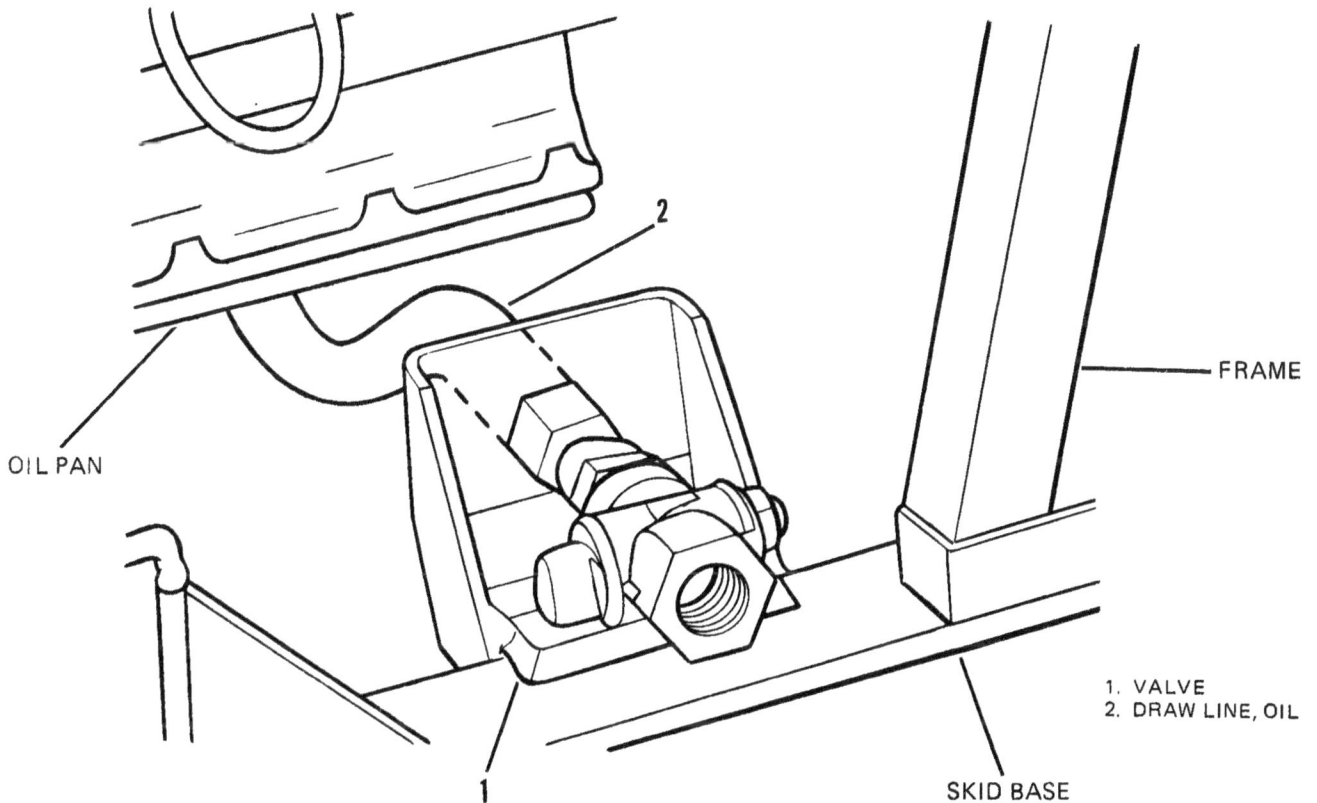

OIL PAN

FRAME

1. VALVE
2. DRAW LINE, OIL

SKID BASE

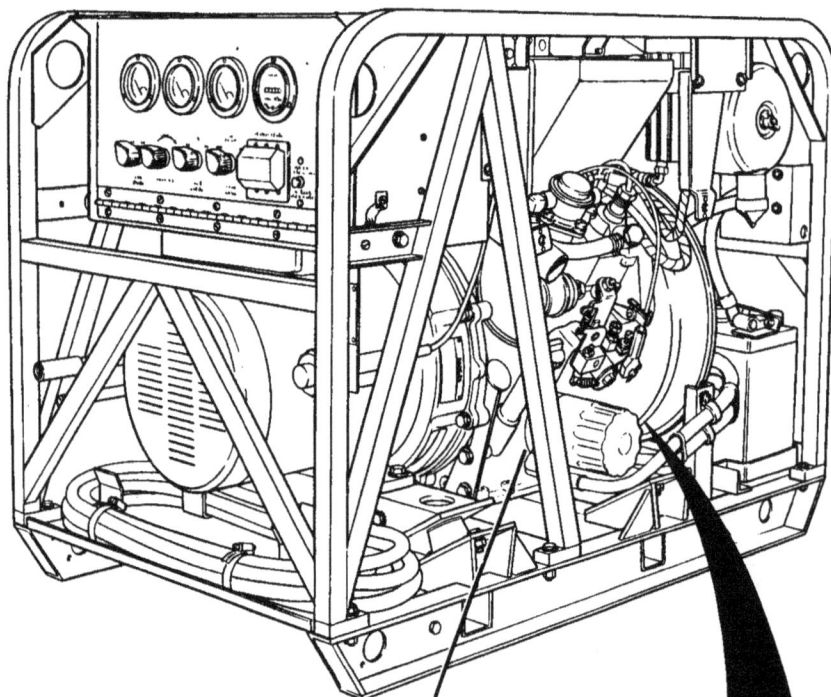

OIL FILTER
ADAPTER

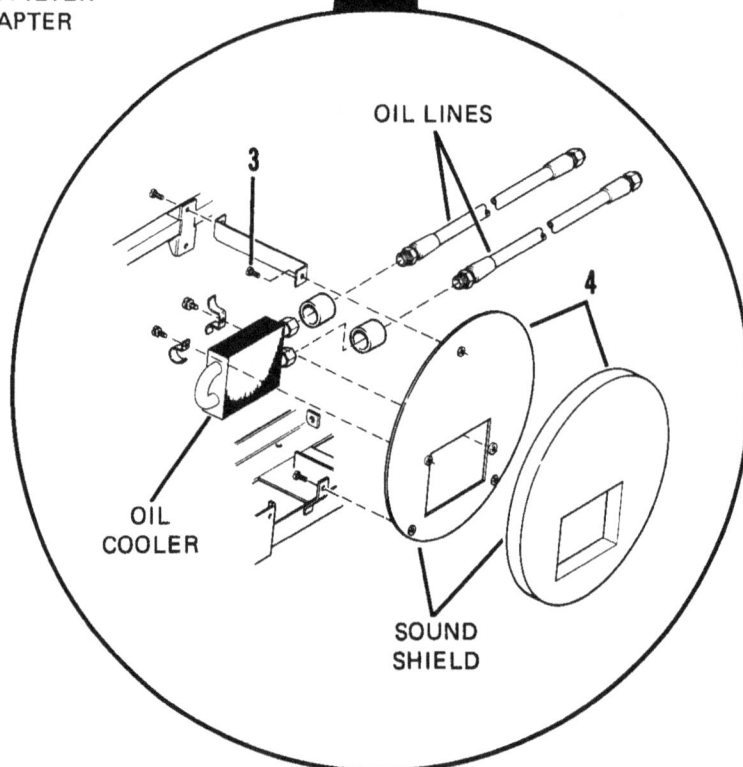

OIL LINES

3

4

OIL
COOLER

SOUND
SHIELD

3. SCREW
4. SHIELD

Figure 2-5. Engine Removal (Sheet 2 of 3).

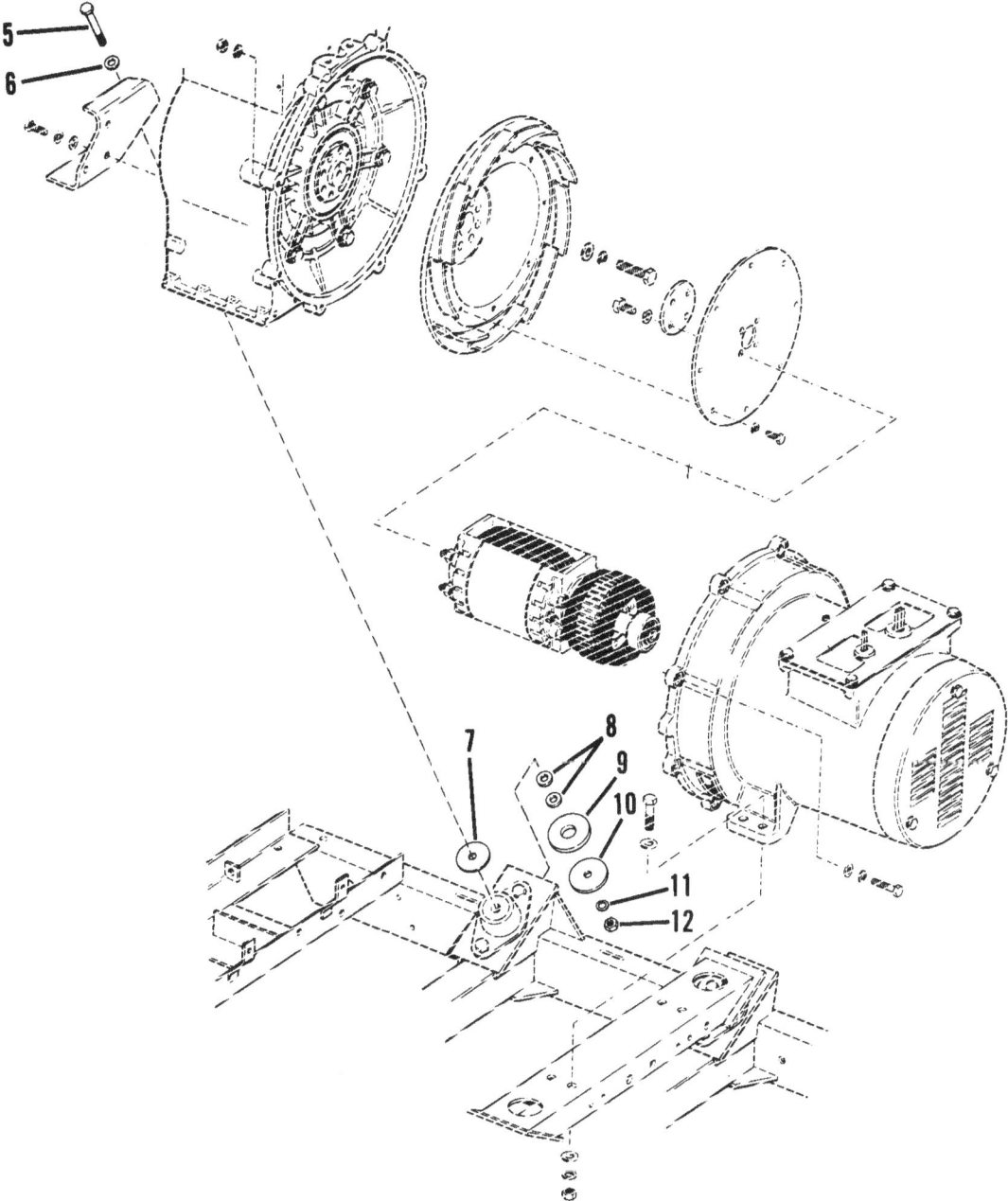

5. SCREW
6. LOCKWASHER
7. WASHER
8. WASHER
9. WASHER
10. WASHER
11. LOCKWASHER
12. NUT

Figure 2-5. Engine Removal (Sheet 3 of 3).

(3)   Removebattery, refer to TM 05926B/06509B-12/TM 5-6115-615 -12/NAVFAC P-8-646-12/TO 35C2-3-386-31 manual paragraph 4-19.

(4)   Remove frame, refer to paragraph 3-2.

(5)   Remove rear starter bracket, refer to TM 05926B/06509B-12/TM 5-6115-615-12/ NAVFAC P-8-646-12/TO 35C2-3-386-31 manual, paragraph 4-48.

(6)   Remove engine oil filter, refer to TM 05926B/06509B-12/TM 5-6115-615-12/ NAVFAC P-8-646-12/TO 35C2-3-386-31 manual, paragraph 4-46.

(7)   Remove screws (3) and sound shield (4).

(8)   Remove oil cooler, refer to paragraph 7-20.

(9)   Disconnect throttle cable from governor, refer 6115-615-12/NAVFAC   to TM 05926B/06509B-12/TM 5-manual,
      P-8-646-12/TO 35C2-3-386-31   paragraph 4-56.

(10) Remove generator, refer to paragraph 2-10.
       fuel pump, refer to TM
(11) Disconnect electrical connector from auxiliary
      05926B/06509B-12/TM 5-6115-615-12/NAVFAC P-8-646-12/TO 35C2-3-386-31 manual, paragraph 4-35.

(12) Attach a chain hoist to the engine lifting eyes.   Remove slack from chains, but do not
      lift engine.

(13) Remove engine wiring harness, refer to TM 05926B/06509B-12/TM 5-6115-615-12/ NAVFAC P-8-646-12/TO 35C2-3-386-31 manual, paragraph 4-72.

(14) Remove air intake hose, refer to TM 05926B/06509B-12/TM 5-6115-615-12/ NAVFAC P-8-646-12/TO 35C2-3-386-31 manual, paragraph 4-49.

(15) Remove fuel lines, refer to TM 05926B/06509B-12/TM 5-6115-615-12/NAVFAC P-8-646-12/TO 35C2-3-386-31 manual, paragraph 4-39.

(16) Restore engine support leg (24, Figure 2-6) to original position.

(17)   Remove screws (5), lockwashers (6), washer (7), washers (8, 9 and 10), lockwasher (11) and nut (12). Keep washers in correct sequence for installation.

      Lift and remove engine with chain hoist.

(18) Remove generator fan, refer to paragraph 5-3.

(19)

(1)   Secure generator fan to engine, refer to paragraph 5-3.

(2)   Place engine in position.

(3)   Install screws (5, Figure 2-5), lockwashers (6), washer (7), washers (8, 9, and 10), lockwasher (11), and nut (12).

MARINE CORPS   TM 05926B/06509B-34
ARMY   TM 5-6115-615-34
NAVY   NAVFAC P-8-646-34
AIR FORCE   TO 35C2-3-386-32

(4)   Install fuel lines, refer to TM 05926B/06509B-12/TM 5-6115-615-12/NAVFAC P-8-646-12/TO 35C2-3-386-31 manual, paragraph 4-39.

(5)   Install air intake hose, refer to TM 05926B/06509B-12/TM 5-6115-615-12/ NAVFAC P-8-646-12/TO 35C2-3-386-31 manual, paragraph 4-49.

(6)   Install engine wiring harness, refer to TM 05926B/06509B-12/TM 5-6115-615-12/NAVFAC P-8-646-12/TO 35C2-3-386-31manual, paragraph 4-72.

(7)   Connect electrical connector to auxiliary fuel pump, refer to TM 05926B/06509B-12/TM 5-6115-615-12/NAVFAC P-8-646-12/TO 35C2-3-386-31 manual, paragraph 4-35.

(8)   Install generator, refer to paragraph 2-10.

(9)   Connect throttle cable to governor, refer to TM 05926B/06509B-12/TM 5-6115-615-12/NAVFAC P-8-646-12/TO 35C2-3-386-31 manual, paragraph 4-56.

(10)   Install oil cooler, refer to paragraph 7-20.

(11)   Position sound shield (4) in frame and secure with screws (3).

(12)   Install engine oil filter, refer to TM 05926B/06509B-12/TM 5-6115-615-12/ NAVFAC P-8-646-12/TO 35C2-3-386-31 manual, paragraph 4-46.

(13)   Install rear starter bracket, refer to TM 05926B/06509B-12/TM 5-6115-615-12/NAVFAC P-8-646-12/TO 35C2-3-386-31 manual, paragraph 4-48.

(14)   Install frame, refer to paragraph 3-2.

(15)   Remove lifting hoist.

(16)   Connect oil drain line engine with (2) to oil pan, close drain valve (1), and fill amount and type of engine oil. the proper TM 05926B/06509B-12/ Refer to 5-6115-615-12/NAVFAC P-8-646-12/TO 35C2-3-386-31 TM

(17)   Install battery, refer to TM 05926B/06509B-12/TM 5-6115-615-12/NAVFAC P-8-646-12/TO 35C2-3-386-31 manual, paragraph 4-22.

NOTE

Generator cannot be removed as an assembly; some disassembly is required.

2-10. GENERATOR.

a. Removal.

(1) Remove frame, refer to paragraph 3-2.

(2) Remove rear starter bracket, refer to TM 05926B/06509B-12/TM 5-6115-615-12/ NAVFAC P-8-646-12/TO 35C2-3-386-31 manual, paragraph 4-48.

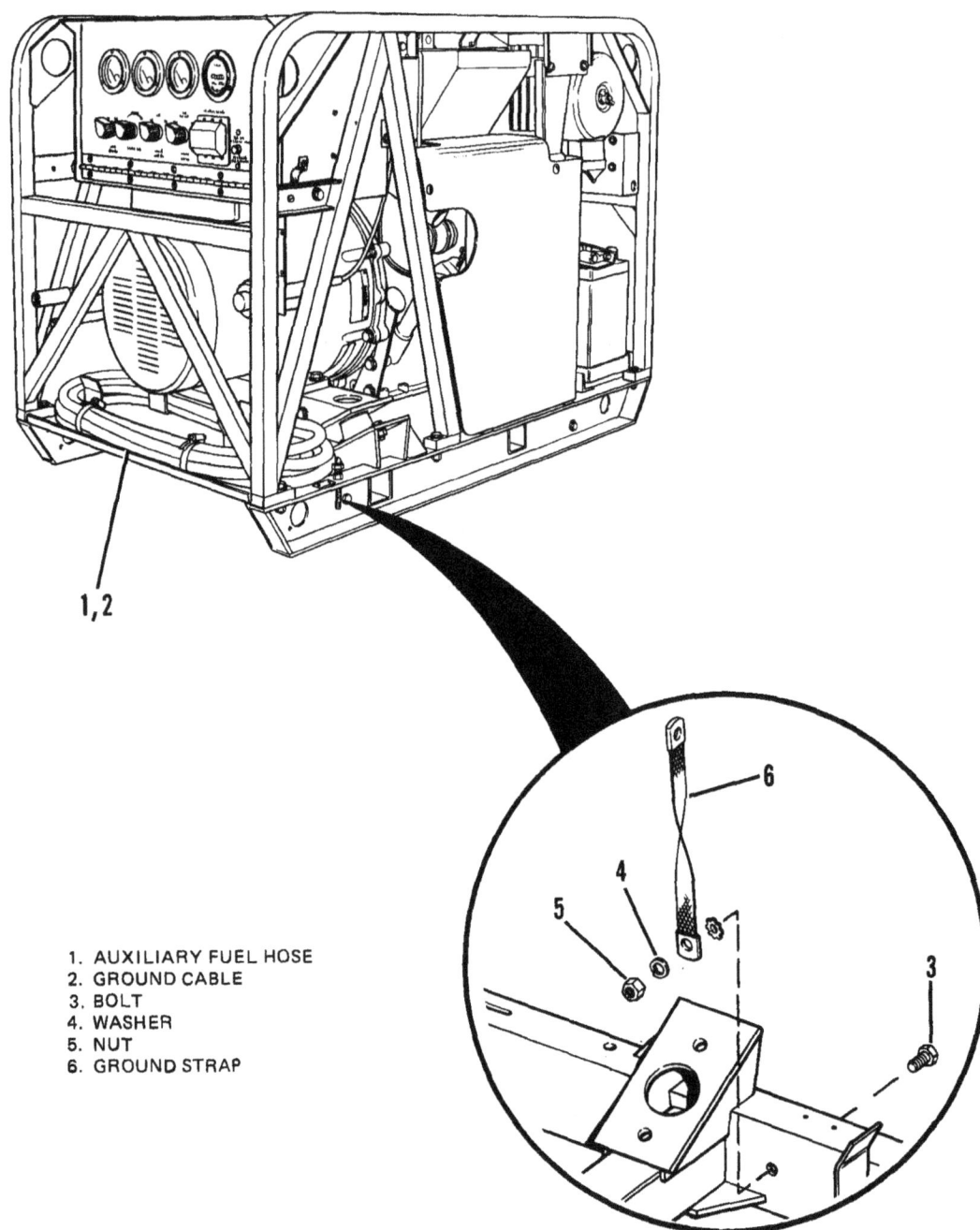

1. AUXILIARY FUEL HOSE
2. GROUND CABLE
3. BOLT
4. WASHER
5. NUT
6. GROUND STRAP

Figure 2-6. Generator Removal and Installation (Sheet 1 of 2).

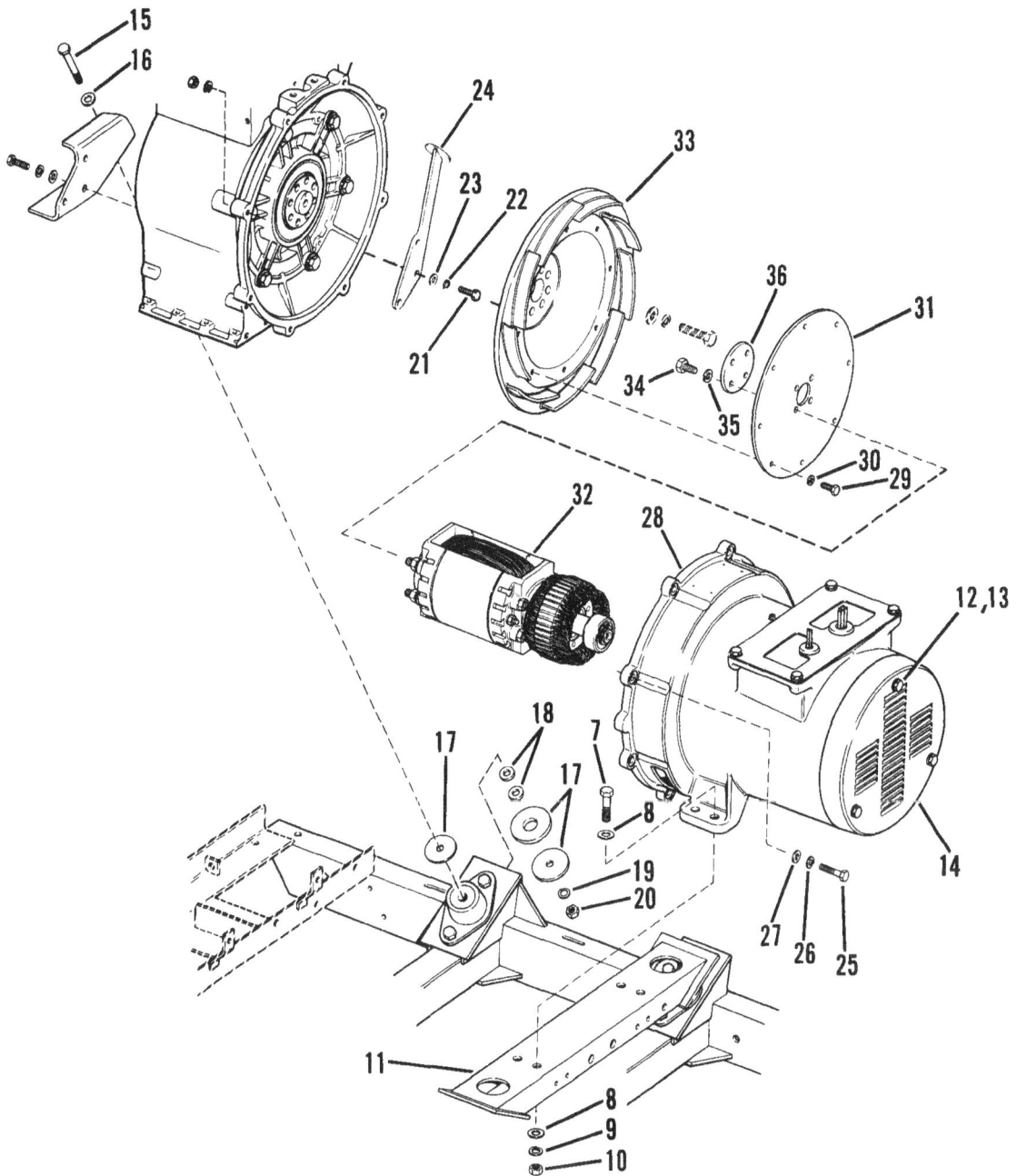

Figure 2-6. Generator Removal and Installation (Sheet 2 of 2).

7. SCREW	17. WASHER	27. WASHER
8. WASHER	18. WASHERS	28. STATOR ASSEMBLY
9. LOCKWASHER	19. LOCKWASHER	29. SCREW
10. NUT	20. NUT	30. LOCKWASHER
11. BRACE	21. SCREW	31. DRIVE ADAPTER
12. BOLT	22. LOCKWASHER	32. ROTOR
13. LOCKWASHER	23. WASHER	33. FAN
14. COVER	24. SUPPORT	34. BOLT
15. SCREW	25. BOLT	35. LOCKWASHER
16. LOCKWASHER	26. LOCKWASHER	36. WASHER PLATE

(3)   Remove control box, refer to paragraph 2-8.

(4)   Attach a hoist to the generator.Remove slack from hoist chains but do not lift generator.

Remove auxiliary fuel hose (1, Figure 2-6) and ground cable (2) from skid base.

(5)

Remove bolt (3), washer (4), and nut (5) securing ground strap (6) to skid base.

(6)   Remove screws (7), washers (8), lockwashers (9), and nuts (10) that secure generator to cross brace (11).

Remove bolts (12), lockwashers (13), and end cover (14) of generator.

(7)

Remove screws (15), lockwashers (16), washers (17), washers (18), lockwashers (19), and nuts (20).Keep washers in correct sequence for installation.

(8)

(9)

(10)   Remove screws (21),   lockwashers (22), washers (23) and engine support (24).

(11)   Turn engine support support   (24) so leg of support is facing down. Leg will generator is removed.
       engine when

                                     engine support (24) to engine with screws (21),

(12)   Install and tighten
       lockwashers (22) and washers (23).

(13)   Remove bolts (25), lockwashers (26), and washers (27) securing generator to engine housings.

(14)   Use a gear puller to remove generator housing and stator assembly (28).

(15)   Remove screws (29) and lockwashers (30) securing drive adapter (31) and rotor (32) to fan (33).

(16)   To remove rotor (32) and drive lightly with a   adapter (31) from fan (33), tap end of rotor Be careful not to tap rotor
       rubber mallet.   windings.

(17)   Remove bolts (34), lockwashers
       (31) from rotor (32).

b.   Installation.

(1)
     Install screws (7), washers (8), lockwashers (9) and nuts (10) that secure generator to cross brace (11).

(2) Install drive adapter (31, Figure 2-6), to rotor (32) with washer plate
     (36), lockwashers (35), and bolts (34).

(3) Put rotor (32) and drive adapter (31) in position on fan (33) and install
     bolts (29), and lockwashers (30).Tighten bolts (29) to 7 to 8 ft lbs (10 to 11 N.m).

CAUTION

Use extreme care when installing stator housing.
Misalignment and/or forced assembly may damage the rotor or stator.

(4) Position stator housing (28) over rotor (32).

(5) Secure generator stator assembly (28) to engine housing with bolts (25), lockwashers (26), and washers (27). Install two long bolts in holes above and below starter motor.

(6) Remove screws (21), lockwashers (22), washers (23) and engine support (24).

(7) Turn engine support (24) so leg of support is facing up.

(8) Install and tighten engine support (24) to engine with screws (21), lockwashers (22) and washers (23).

(9) Install resilient mount screws (15), lockwashers (16), washers (17), washers (18), lockwashers (19) and nuts (20).

(10) Install end cover (14) with bolts (12) and lockwashers (13).

(11) Connect ground strap (6) to cross brace (11) with screw (7), washer (8), and nut (10).

(12) Install frame, refer to paragraph 3-2.

CHAPTER 3

MAINTENANCE OF THE FRAME

3-1. GENERAL.   The frame provides the generator set a point for lifting, protection against physical damage, and a mounting surface for some components, and it is secured to the skid base.The skid base provides a mounting surface for the engine/generator assembly.A battery frame securely holds the battery in an upright position.

3-2. FRAME.

a. Removal.
Disconnect battery cables before removing frame. The high current output of the DC electrical system can cause arcing and/or burns if a short circuit occurs.

WARNING

Remove control box assembly.Refer to paragraph 2-8.

Disconnect two battery cables and slave receptacle cables from battery, remove nuts (1, Figure 3-1) and disconnect negative (2) cable first then disconnect positive cable (3).

(1)   Remove negative slave cable (6) and positive slave cable (7).

(2)
NOTE

Screws (4) are captive and cannot be removed from sound suppression panel (5).

(3)

(4)   Loosen two screws sound   (4) that secure sound suppression panel (5). Remove panel (5).
suppression

lockwasher (9), washer (10), clamp (11), and ground rod
(5)   Remove screw (8), assemblies (12).

(6)   Remove nut (13), screw (14), and clips (15) that secure wiring harness to the muffler heat shield at three locations.

Remove nuts (16), screws (17), and muffler heat shield panel (18) from the frame.
(7)

Tag and disconnect three fuel lines (19) from top of fuel tank (20).   Plug
hose ends and fittings to prevent the entry of contaminants.
(8)

3-1

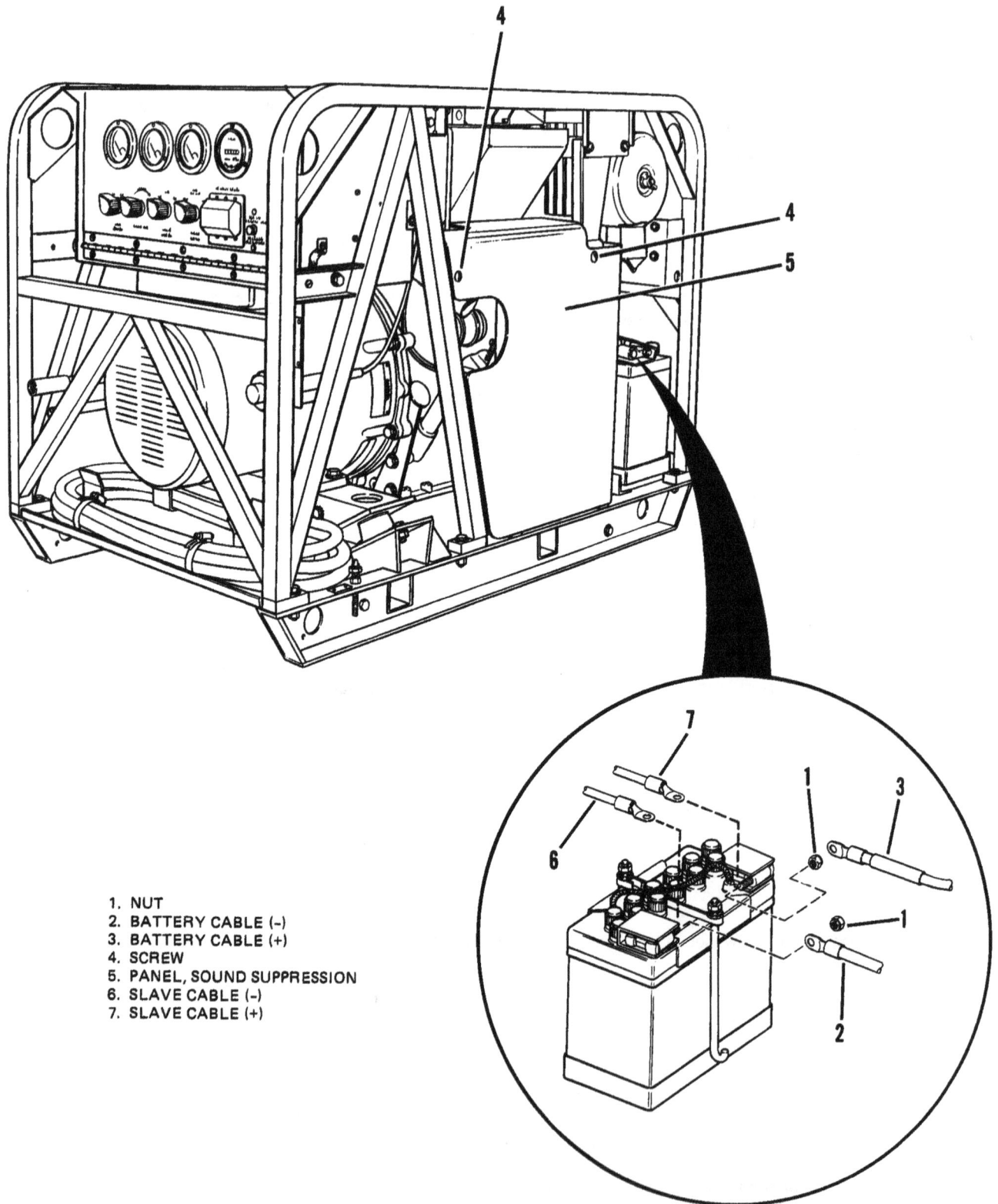

1. NUT
2. BATTERY CABLE (–)
3. BATTERY CABLE (+)
4. SCREW
5. PANEL, SOUND SUPPRESSION
6. SLAVE CABLE (–)
7. SLAVE CABLE (+)

Figure 3-1. Frame Removal (Sheet 1 of 5).

8. SCREW
9. LOCKWASHER
10. WASHER
11. CLAMP
12. GROUND ROD
13. NUT
14. SCREW
15. CLIP
16. NUT
17. SCREW
18. PANEL, HEAT SHIELD

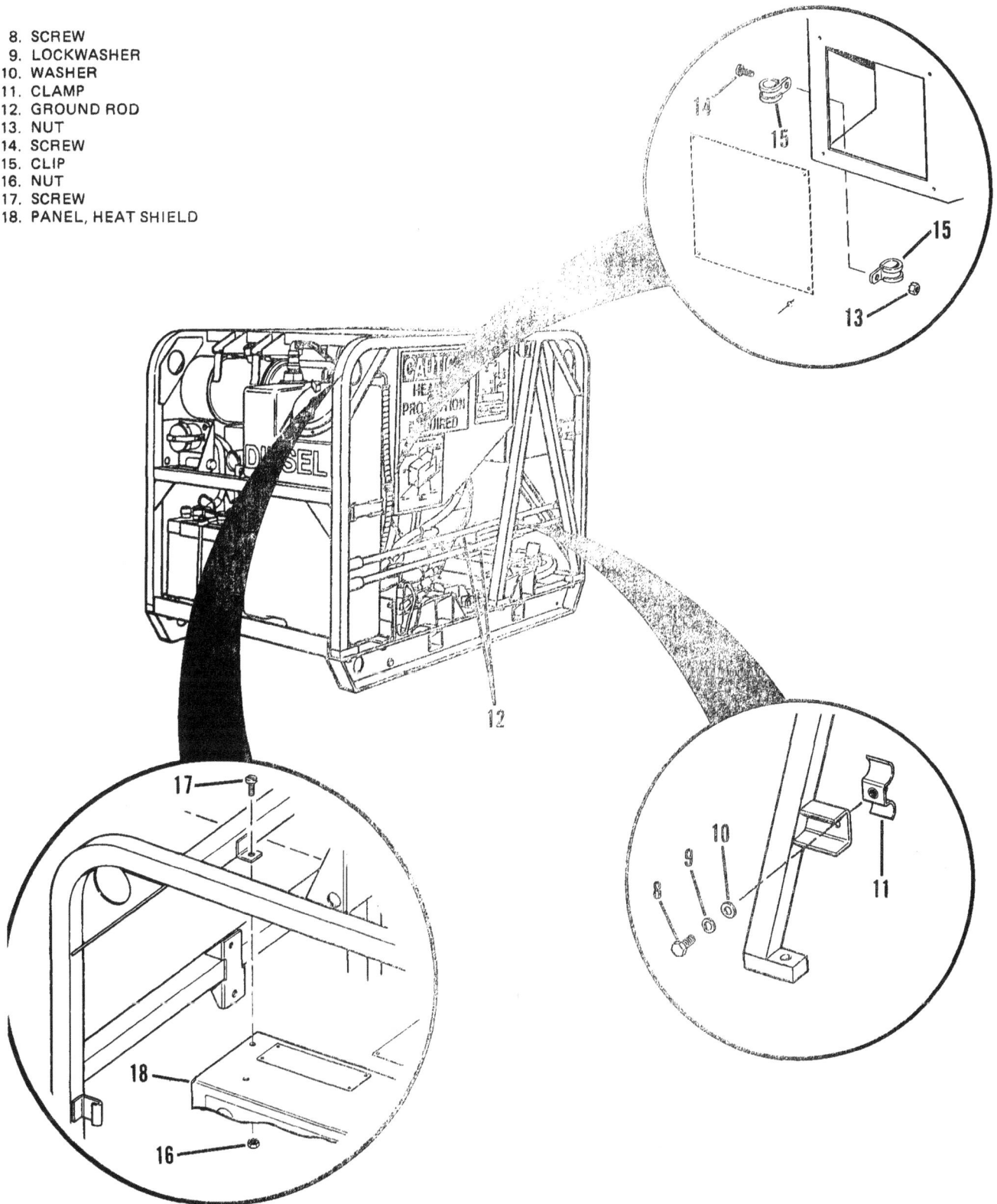

Figure 3-1. Frame Removal (Sheet 2 of 5).

(9)   Unscrew wire connector (21) from top of fuel tank.

(10)   Loosen fuel filter vent plug (22) on fuel filter (23) and drain fuel from filter drain (24) into a suitable container.

(11)   Tag and disconnect two hoses (25 and 26) from fuel filter (23) and plug hose ends and fittings to prevent the entry of contaminants.

(12)   Loosen clamp (27).   Remove four screws (28) lockwashers (29), and washers (30) and remove air cleaner (31).

(13)   Remove nuts (32), lockwashers (33), washers (34), bolts (35), and washers (36) that secure frame to skid base (6 locations).

(14)   Remove screw (37) that secures bracket (38) to frame.

(15)   Remove cap from fuel tank drain.

(16)   Carefully lift frame from skid base, and remove fuel tank (39).

b. Repair.Repair of the frame is limited to simple bending or aluminum

welding.   Refer to paragraph 2-7.6.

c.   Installation.
Install oil cooler, refer to paragraph 7-20.

(1)   Place fuel tank (39) in position on skid base and replace cap from fuel tank drain.

(2)   Carefully place frame in position on skid base.

(3)   Secure fuel tank (39) to frame.

(4)   Secure bracket (38) to frame with screw (37).

(5)   Secure frame to skid base with washers (36), bolts (35), washers (34), lockwashers (33), and nuts (32).

(6)   **NOTE**
                     Make sure rubber top of gen-
                     erator

If control box was removed, push generator wires through grommets in
bottom of box. pad completely surrounds opening in after installation.

(7)   Install control box. Refer to paragraph   2-8.

(8)   Install air cleaner (31)   and secure with   four screws (28), lockwashers
        (29), and washers (30).   Tighten clamp (27).

(9)   Remove plugs and connect   two fuel lines (25 and 26) to fuel filter (23).

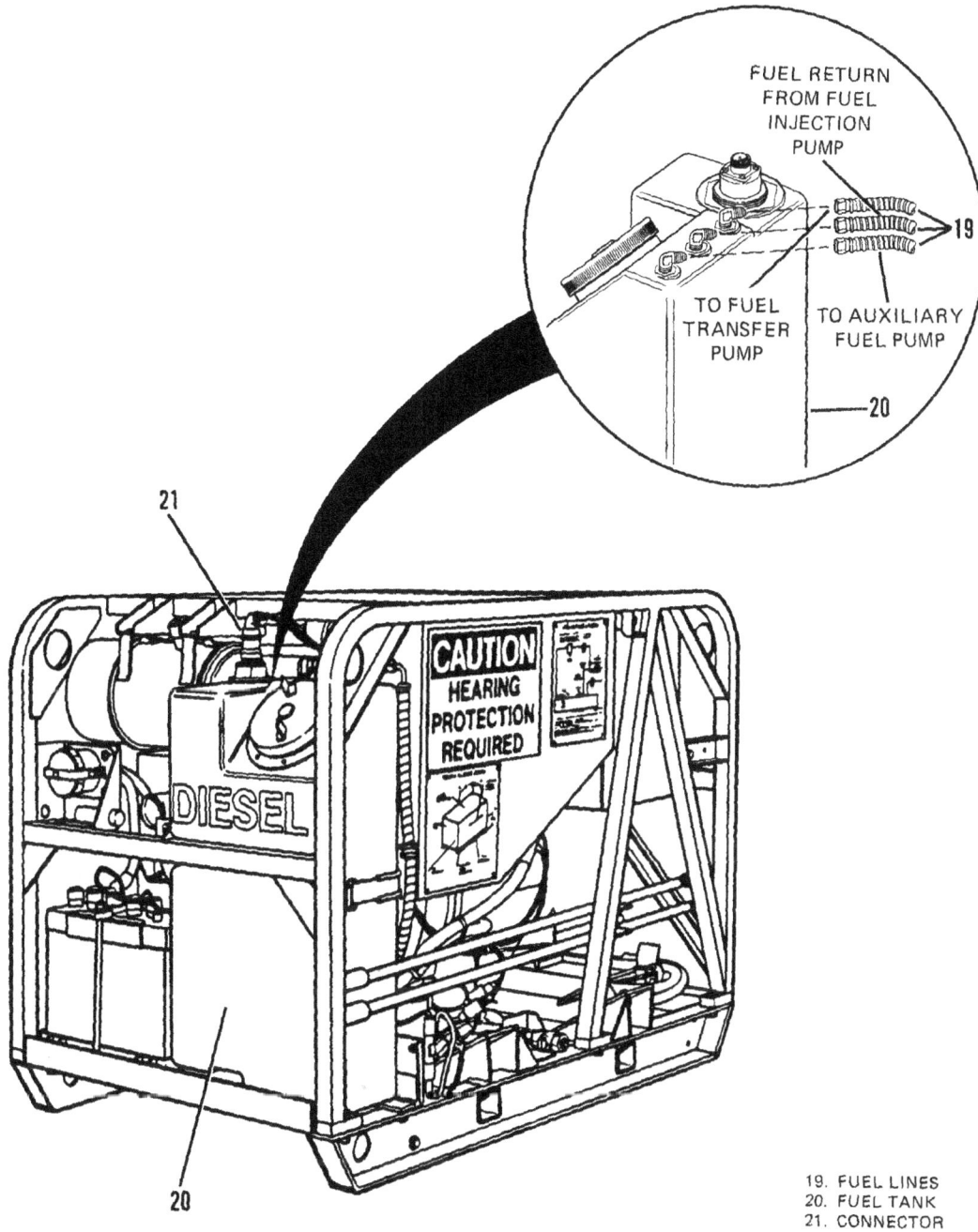

FUEL RETURN
FROM FUEL
INJECTION
PUMP

19

TO FUEL
TRANSFER
PUMP

TO AUXILIARY
FUEL PUMP

20

21

DIESEL

CAUTION
HEARING
PROTECTION
REQUIRED

20

19. FUEL LINES
20. FUEL TANK
21. CONNECTOR

Figure 3-1. Frame Removal (Sheet 3 of 5).

22. PLUG, VENT
23. FILTER, FUEL
24. DRAIN, CONDENSATION
25. HOSE, FUEL
26. HOSE, FUEL
27. CLAMP
28. SCREW
29. LOCKWASHER
30. WASHER
31. AIR CLEANER
32. NUT
33. LOCKWASHER
34. WASHER
35. BOLT
36. WASHER

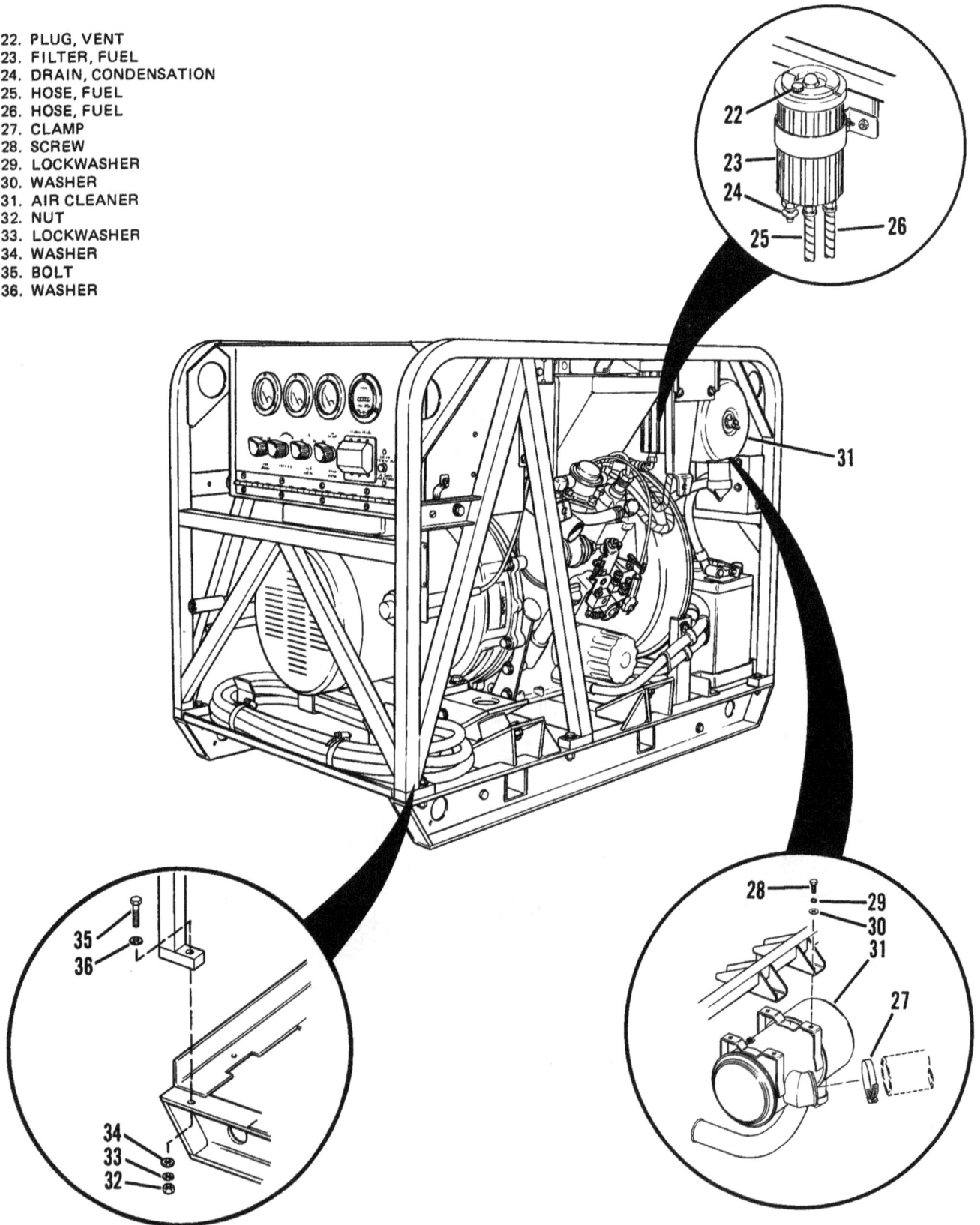

Figure 3-1. Frame Removal (Sheet 4 of 5).

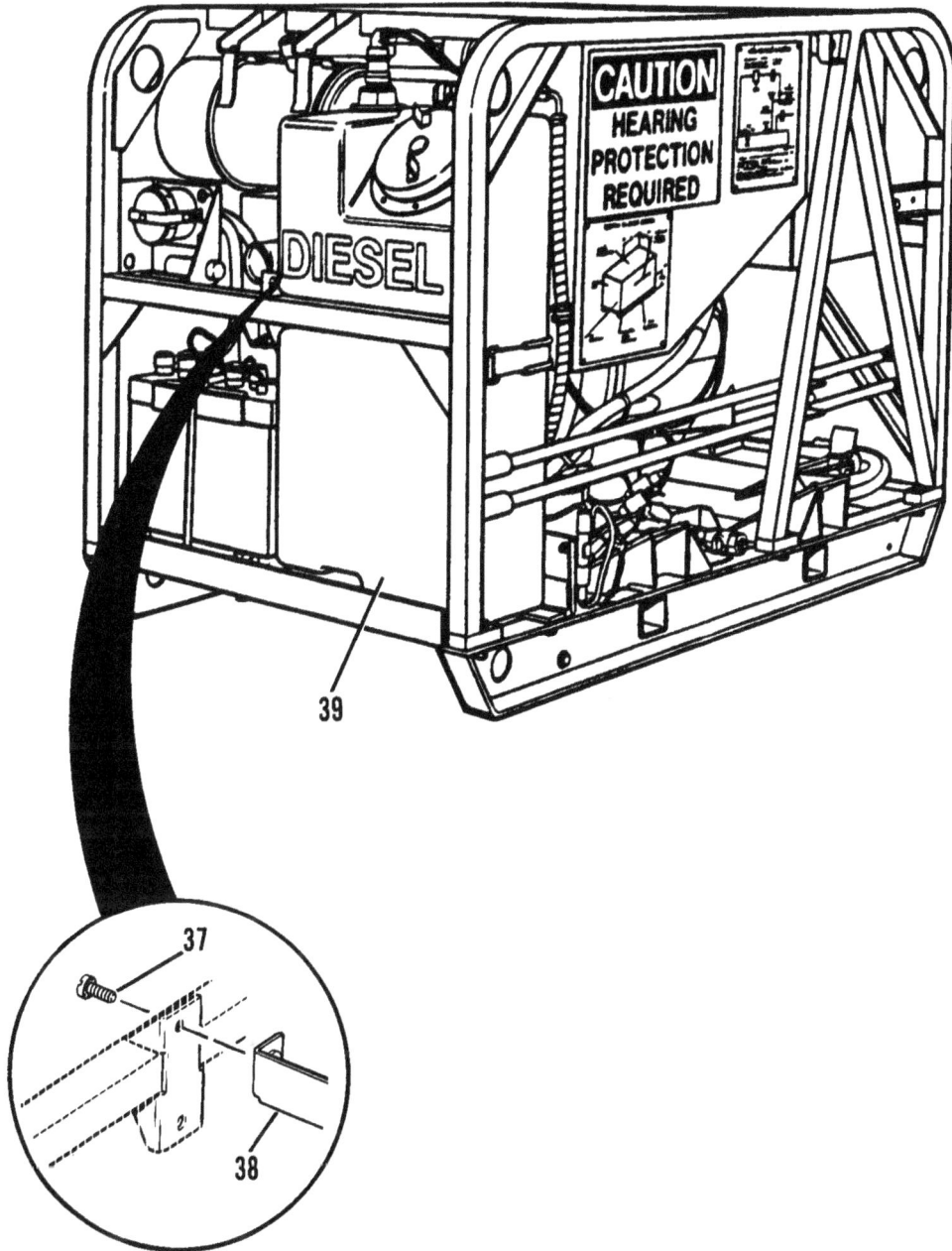

37. SCREW
38. BRACKET
39. FUEL TANK

Figure 3-1. Frame Removal (Sheet 5 of 5).

(10)   Check that vent plug (22) on fuel filter (23) is closed.

(11)   Connect wire connector (21) to top of fuel tank (20).

(12)   Remove plugs from ends of fuel lines (19) and connect to fuel tank (20).

(13)   Place cable clip (15) inside heat shield panel (18).

(14)   Secure heat shield panel (18) to frame with screws (17) and nuts (16).

(15)   Secure wiring harness and auxiliary fuel line with clips (15), screw (14), and nut (13). Clip (15) on outside of heat shield panel secures the auxiliary fuel line and clip (15) inside the heat shield panel secures the wiring harness.

(16)   Secure ground rod assemblies (12) with clamp (11), washer (10), lockwasher (9), and screw (8).

Place sound suppression panel (5) on generator set and tighten screws (4).

(17)   Install battery. Refer to TM 05926B/06509B-12/TM 5-6115-615-12/NAVFAC P-8-646-12/TO 35C2-3-386-31 manual, paragraph3-15.

(18)   Connect battery cables last when installing frame.The high current output of the DC electrical system can cause arcing and/or burns if a short circuit occurs.

WARNING

Connect positive battery cable (3) and positive slave cable (7) first and then negative battery cable (2) and negative slave cable (6) to battery. Secure cables by tightening nuts (1).

3-3.   SKIDBASE.

a. Removal.

(19)

(1) Remove frame, refer to paragraph 3-2.

(2) Remove generator, refer to paragraph 2-10.

(3) Remove engine, refer to paragraph 2-9.

b.Disassembly.

(1) Remove nuts (1, Figure 3-2), lockwashers (2), washers (3), bolts (4) and mounts (5).

(2) Remove nut (6), lockwasher (7), ground strap (8), and bolt (9).

1. NUT
2. LOCKWASHER
3. WASHER
4. BOLT
5. MOUNT
6. NUT
7. LOCKWASHER
8. STRAP, GROUND
9. BOLT
10. NUT
11. LOCKWASHER
12. TERMINAL, GROUNDING
13. PLATE, DATA
14. PLATE, DATA

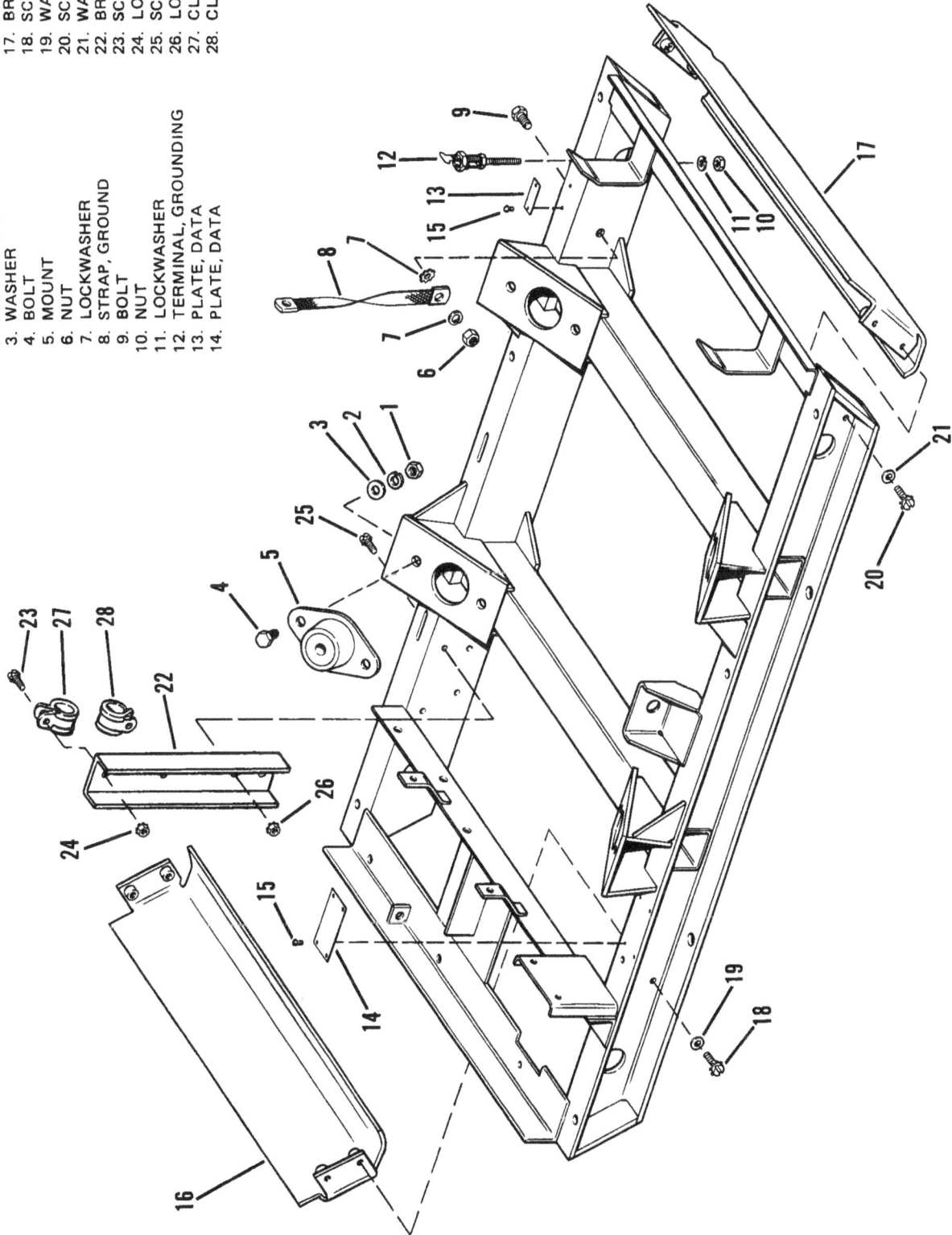

15. RIVET
16. BRACKET, REAR
17. BRACKET, FRONT
18. SCREW
19. WASHER
20. SCREW
21. WASHER
22. BRACKET
23. SCREW
24. LOCKNUT
25. SCREW
26. LOCKNUT
27. CLAMP
28. CLAMP

Figure 3-2. Skid Base.

(3)   Remove nut (10), lockwasher (11), and grounding terminal (12).

(4)   If necessary, remove auxiliary fuel pump, battery tray, and oil drain assembly.   Refer to TM 05926B/06509B-12/TM 5-6115-615-12/NAVFAC P-8-646-12/TO 35C2-3-386-31 manual, paragraphs 4-35, 4-19 and 4-51.

(5)   Data plates (13 and 14) are not removed unless visibly damaged.Plates may be removed by removing rivets (15).

(6)   Remove screws   (18), washers (19), and bracket (16).

(7)   Remove screws   (20), washers (21) and bracket (17).

(8)   Remove screws   (23 and 25), locknuts (24 and 26), clamps (27 and 28) and bracket (22).

c.   Cleaning and Inspection.

(1) Wipe all parts with a clean lint free cloth.

(2) Inspect skid base and other parts for damaged threads, cracks, distortion, or other visible damage.

d.Repair. Repair of the skid base may be accomplished by either bending or

aluminum welding, refer to paragraph 2-7.6. e.   Assembly.

(1) Install bracket (22) with screws (23 and 25), locknuts (24 and 26) and clamps (22 and 28).
Install bracket (17) with screws (20) and washers (21).

(2)   Install bracket (16) with screws (18) and washers (19).

(3)   If removed, data plates (13 and 14) may be secured to skid base with rivets (15).

(4)   If necessary, install oil drain assembly, battery tray, and auxiliary fuel pump. Refer to TM 05926B/06509B-12/TM 5-6115-615-12/NAVFAC P-8-646-12/ TO 35C2-3-386-31 manual, paragraphs 4-35, 4-19 and 4-51.

(5)   Secure grounding terminal (12) to skid base with lockwasher (11)and nut (10).

(6)   Secure ground strap (8) to frame with bolt (9), lockwashers (7) and nut (6).

(7)   Install mounts (5), bolts (4), washers (3), lockwashers (2), and nuts (1).

(8)

f.   Installation.

(1) Install engine, refer to paragraph 2-9.

(2) Install generator, refer to paragraph 2-10.

(3) Install frame, refer to paragraph 3-2.

CHAPTER 4

MAINTENANCE OF THE DC ELETRICAL AND CONTROL SYSTEM

4-1. GENERAL. A battery charging voltage regulator provides a stable DC current that is used to charge the battery after starting. The battery charging voltage regulator is a sealed unit and cannot be repaired or adjusted.

4-2. BATTERY REPAIR.

    a. <u>Battery Case.</u>

        (1) Cracks on the top of the battery case can be repaired. Cracks on the sides or bottom of the battery cannot be repaired and the battery should be discarded in accordance with local policy and procedure.
        (2) Cracks in the battery top may be repaired by using epoxy kit (Table 2-2,

    item 1). Reference instructions are furnished with each kit.

    b.   <u>Battery Posts.</u>

        (1) Battery posts which are worn or broken may be rebuilt by using pig lead and steel battery post molds. Pig lead is available in five pound bars (Table 2-2, item 2). Post molds are supplied in two sizes: positive (+) post, NSN 5120-00-251-5045 and negative (-) post, NSN 5120-00-251-5046. Positive post mold has a larger inside diameter than the negative post mold.

WARNING

Before repairing battery posts, place battery on a work bench under an exhaust hood to protect personnel from lead and/or acid fumes.

WARNING

Always wear safety goggles and gloves when repairing battery posts.

WARNING

Follow procedures (2) through (5) carefully to avoid danger of explosion.

        (2) Remove all filler caps from the battery.

(3) Make sure that electrolyte level is up to ledge in filler openings. Add distilled water as necessary. A low electrolyte level creates a potential for hydrogen gas to be trapped in the plate area. Hydrogen gas when subjected to flame or sparks can explode.

(4) Let the battery stand for five minutes to permit explosive hydrogen gas to dissipate.

(5) Using a low flame from a torch, flash the area above the open cells to remove excess fumes. There may be a small flash of flame when this is done, so stand clear of the immediate area over the battery.

Clean the battery post. Dress down post to remove all burrs, rough edges, and corrosion.

(6) Place the battery post mold securely around the post to be repaired. Seat post mold to prevent molten lead from running out from under the mold.

(7) With steel post mold securely in place, apply heat to the top of the damaged post. Use a torch that produces a narrow pencil shaped flame so that heat can be applied to battery post without overheating the post mold. An overheated post mold will cause the lead to become too hot and lead will run out of mold base.

(8)

(9) Once the lead starts to melt, do not jar and unseat the mold.

(10) When the top of lead post is molten, hold one end of the lead stick close over the mold, heat the end of the lead stick, and feed the molten lead into the already molten metal of the battery post.

NOTE

To prevent separation of old and new lead, do not jar the battery while the lead is solidifying.

(11) When lead has hardened and cooled, the mold can be removed by working off the newly formed post, slowly, with a pair of pliers.

(12) Test the newly formed post by gripping top of post with a pliers and twisting. If top breaks off, the lead was not fully molten to effect a complete bonding of the old and new lead, and the repair procedure must be repeated.

(13) If the post was properly rebuilt, it will roughness. Dress and clean the post with

4-2

CHAPTER 5

MAINTENANCE OF THE ELECTRICAL POWER GENERATION AND CONTROL SYSTEM

5-1. GENERAL. Repair of the generator is limited to the replacement of defective parts.   The generators on each of the models covered by this manual are nearly identical in appearance. Testing procedures are similar but may yield different results depending on the generator model.

5-2. GENERATOR ASSEMBLY.

a. Testing.

NOTE

Before testing power generator make sure that the voltmeter on control panel is operating properly, refer to procedures contained in operator/organizational manual. Also make sure that the output voltage

regulator is functioning properly, refer 5-9.   to paragraph

Start generator set and allow it to warm up.

(1)

Observe the voltmeter on the front of control regulator, or stator is
(2) defective; there will reading on the set voltmeter.   box.   If exciter, rotor, be no or a

(3) Perform the following preliminary generator tests to determine which component of the power generator is faulty. If these tests do not indicate a faulty component, disassembly and testing of individual components is required.

Test exciter stator windings.

(4)

(a) Disconnect   the two exciter stator windings at the control box.

ı____Models   MEP-016B and MEP-021B (60 and 400Hz) disconnect exciter leads F1 and F2 at
stator A1- terminals #1 and #2 of terminal block
(see Figure 5-l).

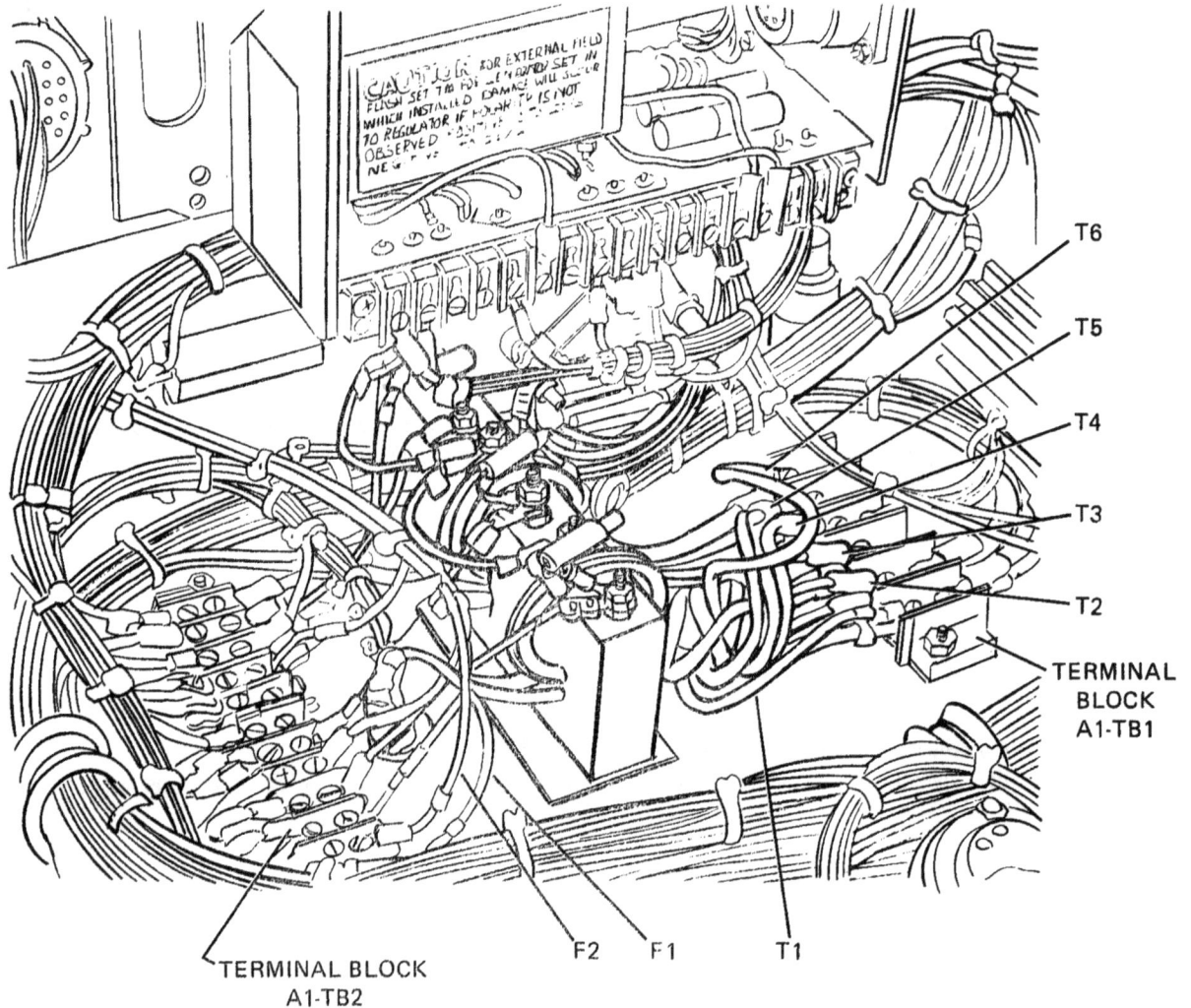

Figure 5-1. Generator Testing-MEP-016B and MEP-021B.

terminals #1 and #2 of terminal board A3-TB1 (see Figure 5-2).

(b)   Connect a multimeter (set to read ohms) across the two exciter winding leads (leads F1 and F2 for models MEP-016B and MEP-021B).

(c)   A reading of approximately 22 ohms indicates that the exciter stator windings are functional.

A low reading of less than 20 ohms indicates shorted windings, and exciter stator should be replaced.

(d)   A reading higher than 25 ohms indicatespoor connections or open windings, and exciter stator should be replaced.

(e)

TERMINAL
BOARD
A3-TB1

EXCITER
STATOR LEAD

EXCITER
STATOR LEAD

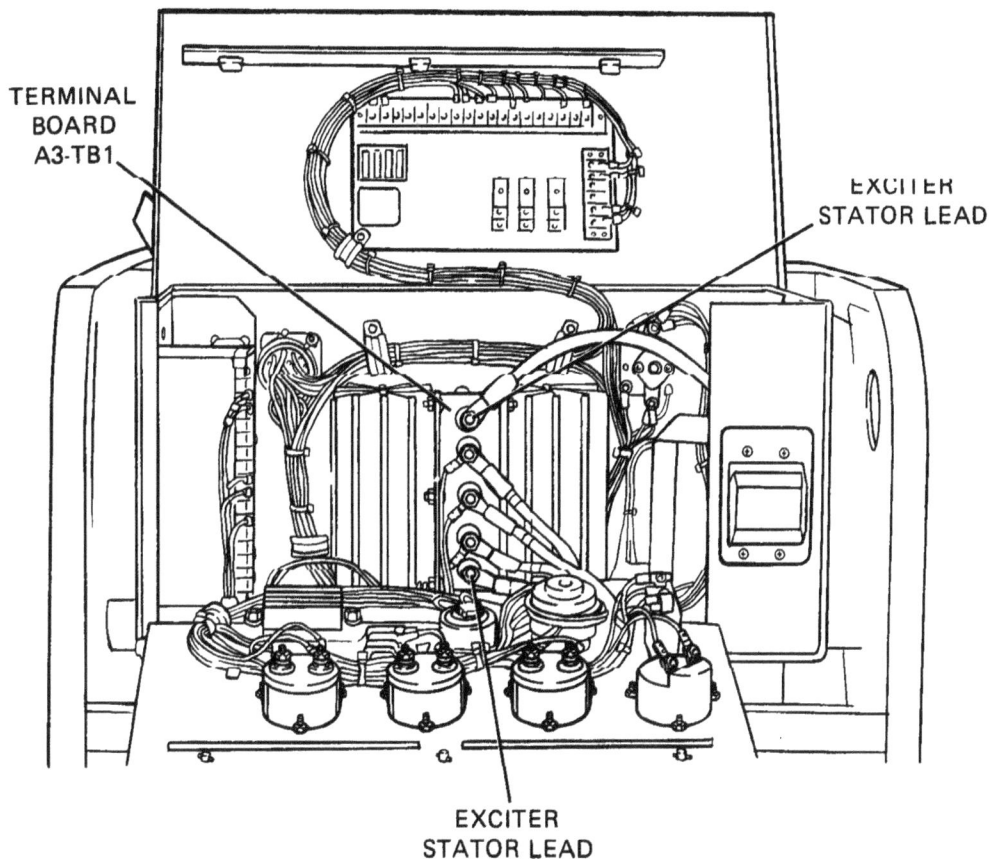

Figure 5-2.Generator Testing-MEP-026B.

(5) Test main stator windings.

    (a) Tag and disconnect the main   1 for MEP-016B and MEP-021B

1   MEP-016B and MEP-021B (60 and 400 Hz) sets disconnect the six stator leads (T1, T2, T3, T4, T5, and T6) from the control box. Lead T1 is connected to terminal #1 of terminal block A1-TB1 (see Figure 5-1).   Lead T2 is connected to terminal #2 of A1-TB1, lead T3 is connected to terminal #3 of A1-TB1, lead T4 is connected to terminal #4 of A1-TB1, lead T5 is connected to terminal #5 of A1-TB1, and lead T6 is connected to terminal #6 of A1-TB1.

        MEP-026B (28 VDC) set disconnect the three stator leads (T1, T2, and T3) from the control box. Lead T1 is connected to terminal #4 of terminal block A3-TB1 (see, Figure 5-2).Lead T2 is connected

2   to terminal #2 of A3-TB1 and lead T3 is connected to terminal #3 of A3-TB1.

    (b) Use a multimeter (set to read ohms) to check the resistance between stator lead pairs.Refer to 1 for MEP-016B and MEP-021B sets and refer to 2 for MEP-026B set.

<u>1</u>____MEP-016B and MEP-021B (60 and 400 Hz) sets check the resistance
between leads (T3 and T6), (T1 and T4), and (T2 and T5).

<u>2</u>____MEP-026B set check the resistance between lead pairs (T1 and T2),
(T2 and T3), and (T1 and T3).

(c)   The stator resistance values should be within the following ranges. If resistance is less than stated; stator has a short and should be replaced.   If resistance is greater than stated; stator has poor connections or open windings and should be replaced.

<u>1</u>____MEP-016B set- the
should be between | main stator winding resistance between lead pairs 0.304 and 0.370 ohms.

<u>2</u>   MEP-021B set- the should
be between | main stator winding resistance between lead pairs 0.149 and 0.181 ohms.

main stator winding resistance between leads 0.018 and 0.022 ohms.

<u>3</u>____MEP-026B set- the
should be between | exciter) windings cannot be tested unless rotor

(3) Generator rotor (main and

is removed from the generator housing, refer to paragraph 5-4.

b.   <u>Removal.</u>____For removal procedures, refer to paragraph 2-10.

c. <u>Disassembly.</u>

NOTE

Figure 5-3 illustrates the MEP-016B and MEP-026B (60Hz and 28VDC) sets.Model MEP-021B (400 Hz) set is similar, but has a slightly different rotor. Procedures which follow can be used for all three models.

(1)   Remove four screws (1, drive   Figure 5-3), lockwashers (2), washer plate (3), and rotor assembly.
adapter (4) from

use a puller to remove bearing (6).
(2)

(3)   Exciter rotor (7) is permanently attached to main rotor (8) shaft and cannot be removed without damaging rotor.

(4)   Remove four screws (9), lockwashers (10) and cover (11).   Remove small

(5)   Remove screw (14), lockwasher (15), and two   bearing retainers (16).

(6)   Data plate (17) is not removed unless it is retained by rivets (18).   being replaced. Data plate is

MARINE CORPS TM 05926B/06509B-34
ARMY   TM 5-6115-615-34
NAVY   NAVFAC P-8-646-34
AIR FORCE   TO 35C2-3-386-32

1. SCREWS
2. LOCKWASHER
3. WASHER PLATE
4. DRIVE ADAPTER
5. CIRCLIP
6. BEARING
7. EXCITER ROTOR
8. MAIN ROTOR
9. SCREWS
10. LOCKWASHERS
11. COVER
12. GROMMET
13. GROMMET
14. SCREW
15. LOCKWASHER
16. BEARING RETAINERS
17. DATA PLATE
18. RIVETS

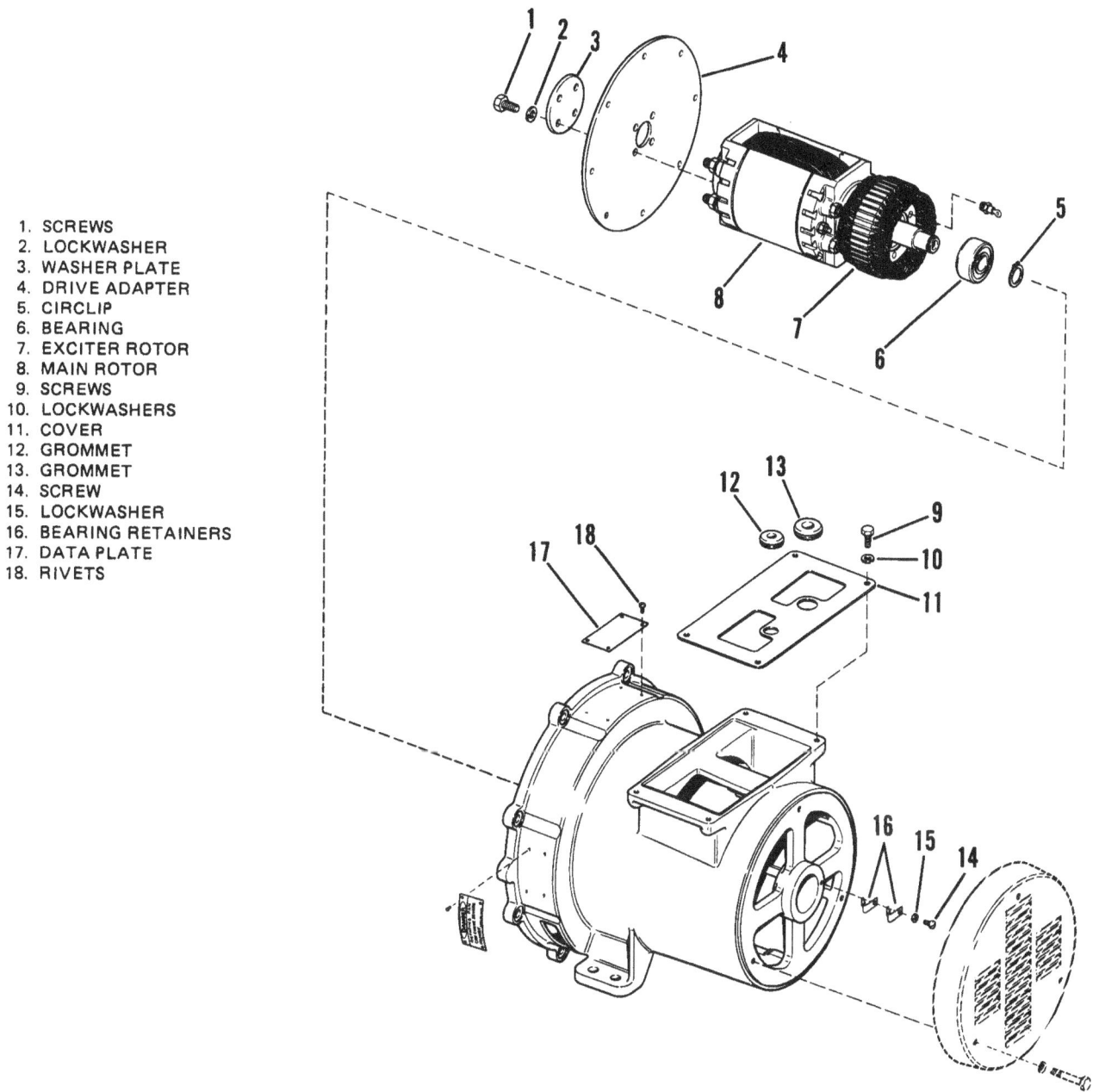

Figure 5-3. Generator Disassembly and Assembly.

(1)   Clean non-electrical parts in solvent and dry with compressed air.

(2)   Wipe all electrical components with a clean, dry cloth.   Clean rotor and stator
      windings with compressed air.

(3)   Inspect rotor and stator windings for signs of burning or damaged insulation.

(4) Inspect rotor and stators for scoring or other indications that stators and rotor have been rubbing together.

(5) Inspect stator leads for damaged or missing insulation.

e. Repair.   Repair of generator is accomplished by replacement of defective components.

f. Assembly.

(1) If removed, install data plate (17, Figure 5-3) with rivets (18).

(2) Secure bearing retainers (16) to housing with screw (14) and lockwasher (15).

(3) Place   large grommet (13) and small grommet (12) into holes in cover (11).

(4) Place   two exciter stator leads through small grommet and six main stator
     leads   (MEP-016B and MEP-021B) or three main stator leads (MEP-026B) through grommet.
     large   Secure cover (11) with four screws (9) and lockwashers (10).

(5) Press bearing (6) onto rotor shaft and install circlip (5).

(6) Install drive adapter (4) and secure with washer plate (3), lockwashers
     (2), and screws (1). Make sure that radius on washer plate (3) faces drive adapter (4). Tighten screws (1) to 25 to 29 ft
(7) lbs (34 to 39 N.m).

g.   Installation.   For installation procedures, refer to paragraph 2-10.

5-3. GENERATOR FAN.

a. Removal.

(1) Remove generator, refer to paragraph 2-10.

(2) Remove screws (1, Figure 5-4), lockwashers (2) and washer (3).

(3) Remove fan (4) from engine crankshaft.

b. Cleaning and Inspection.

(1) Wipe fan with a clean lint-free cloth that has been slightly   moistened with solvent.

(2) Inspect fan for cracks, distortion, or other visible damage.

(3) If fan is cracked or damaged, fan should be replaced.

c. Installation.

(1) Secure fan (4) to engine crankshaft with washers (3), lockwashers (2) and screws (1).   Tighten screws (1) to 51 to 55 ft lbs (69 to 75 N.m).

ENGINE

CRANKSHAFT

1. SCREW
2. LOCKWASHER
3. WASHER
4. FAN

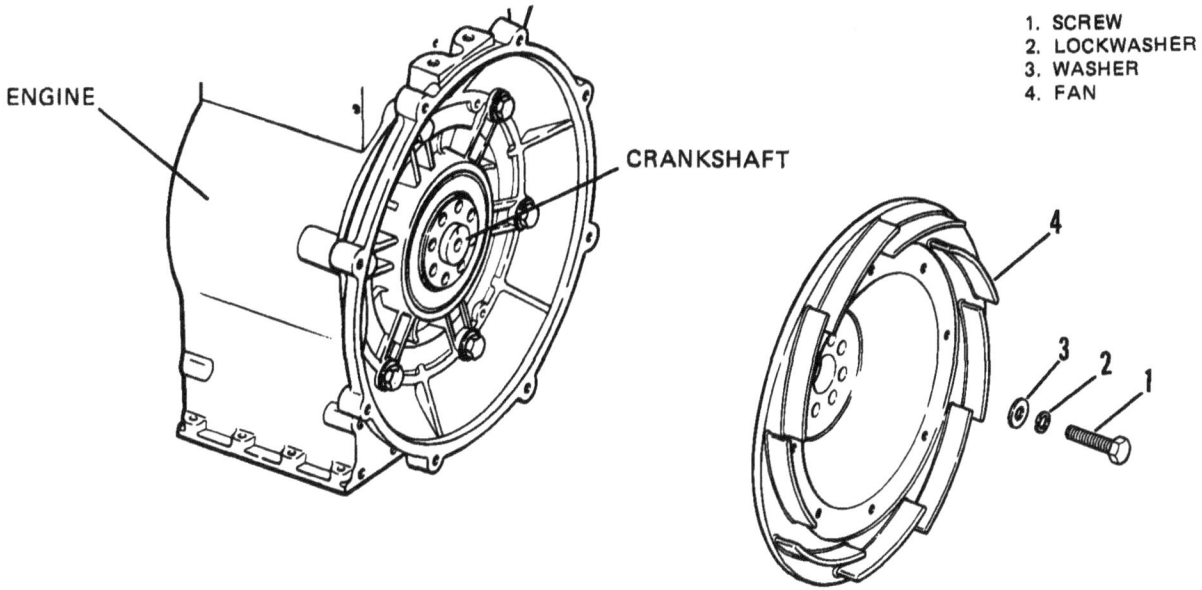

Figure 5-4.Generator Fan.

(2) Install generator, refer to paragraph 2-10.

5-4. ROTOR.

a. Removal.

(1) Remove generator housing and stator assembly, refer to paragraph 2-10.

(2) Remove screws (1, Figure 5-5), lockwashers (2) and washer plate (3).

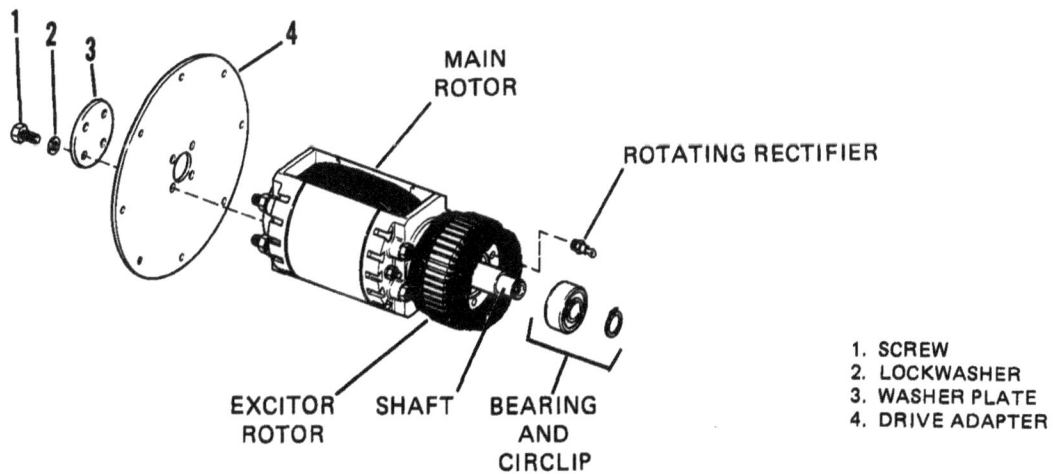

MAIN
ROTOR

ROTATING RECTIFIER

EXCITOR
ROTOR   SHAFT   BEARING
AND
CIRCLIP

1. SCREW
2. LOCKWASHER
3. WASHER PLATE
4. DRIVE ADAPTER

Figure 5-5.Rotor.

(3) Remove drive adapter (4).

CAUTION

DO not remove exciter rotor from rotor shaft.

(5) Remove bearing and circlip from rotor shaft, refer to paragraph 5-8.

b. Testing.

NOTE

Before testing, remove rotating rectifiers (refer to paragraph 5-5) and disconnect main rotor winding lead
(2, Figure 5-8) from exciter core.

(1) Perform high potential test on rotors (exciter and generator).

WARNING

Observe safety regulations. The voltages used in this test are dangerous to human life. Contact with the leads or the windings under test may cause severe and, possibly, fatal shock.Arrange the high voltage leads so that they are not in a position to be accidentally touched. Keep clear of all energized parts.Always reduce the test voltage to zero and ground the winding under test before making any mechanical or electrical adjustments on the equipment.When grounding out windings which have been tested, always connect the connection wire to ground first, and then to the winding.Never perform this test without at least one other person assisting. Generator frame shall be securely grounded.

General.The high potential test is performed to determine whether or not the insulation of the equipment under test is defective.   It is customary to determine whether electrical equipment will withstand normal voltage stresses by means of a test in which higher voltages than normal are applied for a definite period of time. The applied voltage must not be so high (above 1,500 V) as to damage the insulation unless the insulation was initially defective.
(a)

Test Equipment.Use acceptable high potential test equipment in accordance with instructions accompanying the equipment.   Refer to Table 2-1 or Appendix B, Section III of TM 05926B/06509B-12/TM 5-6115-615-12/NAVFAC P-8-646-12/TO 35C2-3-386-31 manual.
(b)

(c)Procedure.

$\underline{1}$ Adequately ground high potential test equipment (1, Figure 5-6) to
a water pipe or similar electrical ground in accordance with instructions accompanying the test
equipment.

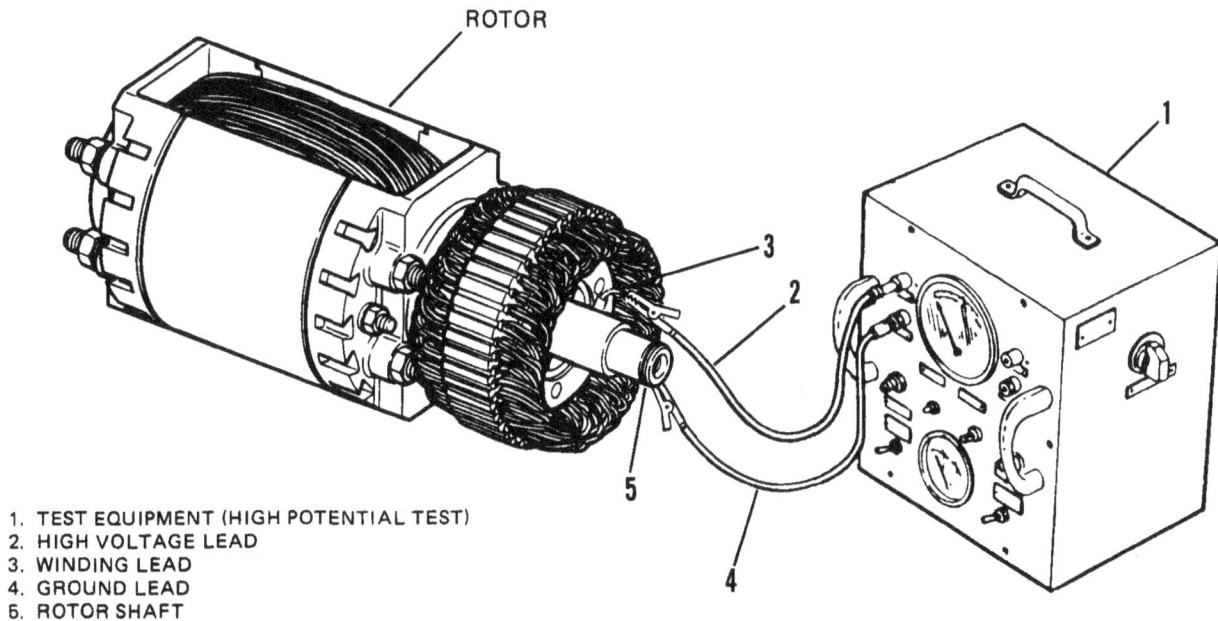

ROTOR

1. TEST EQUIPMENT (HIGH POTENTIAL TEST)
2. HIGH VOLTAGE LEAD
3. WINDING LEAD
4. GROUND LEAD
5. ROTOR SHAFT

Figure 5-6.Rotor Insulation Test.

CAUTION

Make sure that the rotating rectifiers have been removed from the rotor prior to
testing (refer to paragraph 5-5).   Damage to rotating rectifiers may result if high
voltage is applied.

$\underline{2}$ Disconnect ground lead to exciter core (see Figure 5-7).

$\underline{3}$ Connect the high voltage lead (2, Figure 5-6) of the test equipment
to the winding (3) under test.

$\underline{4}$ Connect the ground lead (4) (or grounding bed) to the rotor shaft
(5).

Test.

(d) $\underline{1}$ Turn on test equipment in accordance with manufacturer's instructions after making sure that the initially
applied voltage will be not greater than 600 volts.

<u>2</u>   The test voltage then shall be raised uniformly to the maximum (1,500 Volts RMS, required be 60 Hz). This increase shall accom- 30 seconds. plished in not less than 10 seconds nor more than

<u>3</u>   Apply the maximum voltage for 1 minute.

then shall

<u>4</u>

After 1 minute of applied maximum voltage, the voltage be reduced gradually to the voltage initially supplied. This reduction shall not be accomplished in less than 5 seconds.

<u>5</u>

CAUTION

Ground the high voltage lead of the make sure that no test equipment to the windings charge remains on have been under test. which

<u>6</u>   Remove the high-voltage lead (2)

with the tests of the remaining circuits to be tested. Make sure that all circuits not under test are securely grounded.

(e) Results.   Any evidence of insulation breakdown is cause for replacement of the equipment under test.

(2) Test winding resistance.

(a) Perform winding resistance test using a multimeter (set for ohms).

(b) Check the main rotor windings by connecting the multimeter across main rotor leads (1 and 2, Figure 5-7). Use sharp multimeter probes to penetrate wire insulating material.

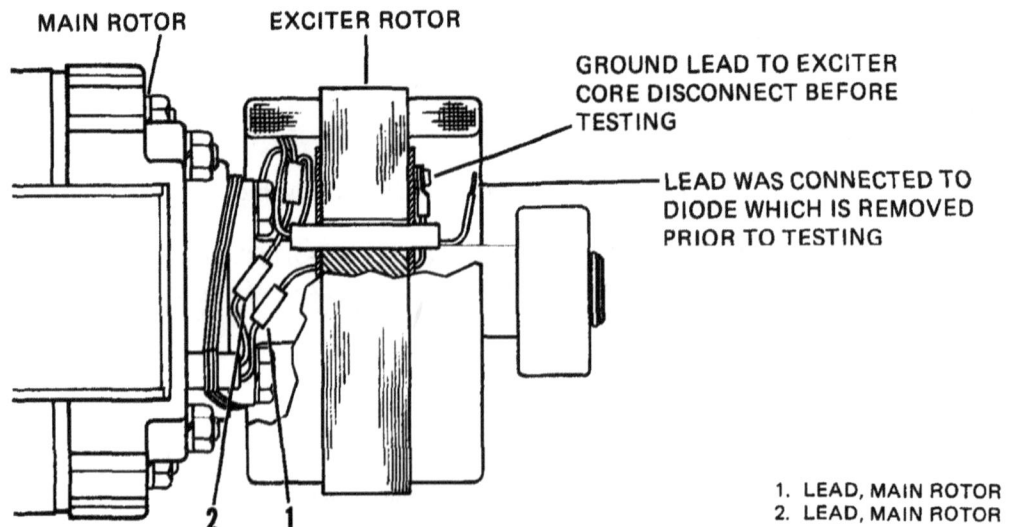

Figure 5-7. **Main Rotor Winding Resistance Check.**

(d) Resistance for ranges: the   main rotor windings should be within the following

MEP-016B
windings   and   MEP-026B (60 Hz and 28VDC) sets: Resistance of rotor at

MEP-021B

should be between 5.5 and 6.7 ohms.

(e) Check the exciter windings by connecting the multimeter across the exciter rotor lead pairs (1,

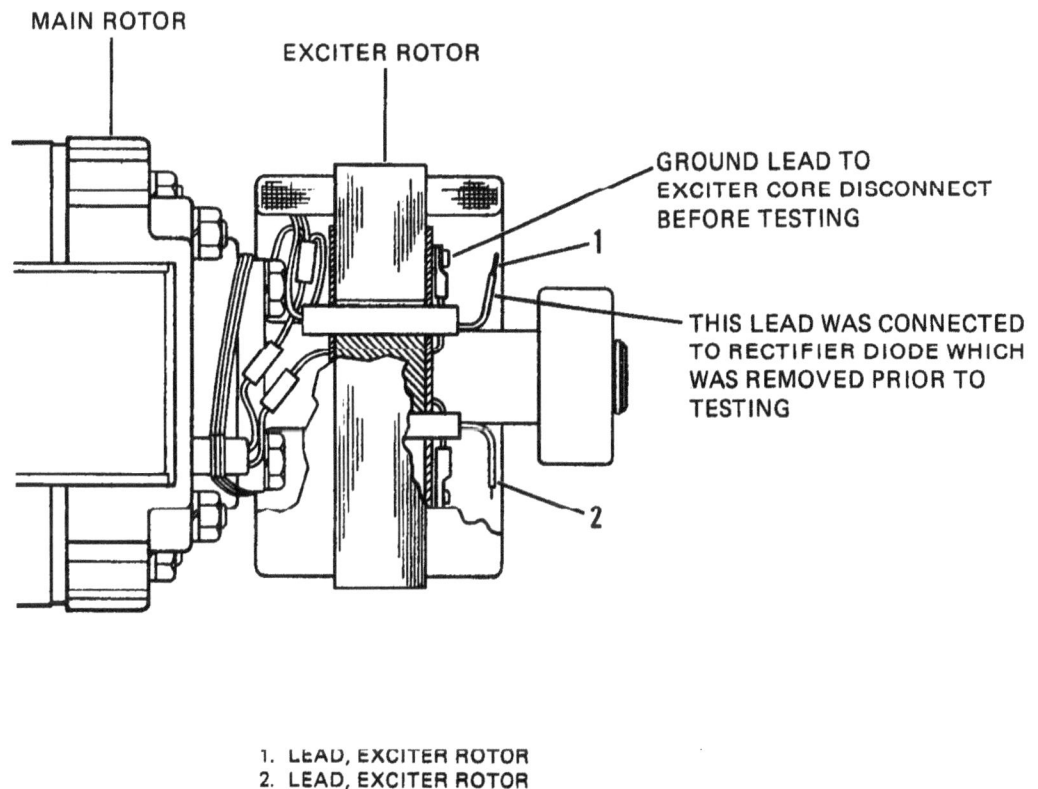

MAIN ROTOR

EXCITER ROTOR

GROUND LEAD TO
EXCITER CORE DISCONNECT
BEFORE TESTING

1

THIS LEAD WAS CONNECTED
TO RECTIFIER DIODE WHICH
WAS REMOVED PRIOR TO
TESTING

2

1. LEAD, EXCITER ROTOR
2. LEAD, EXCITER ROTOR

Figure 5-8), and (2).
Figure 5-8. Exciter Rotor Winding Resistance Check.

(f)
Resistance between lead pairs at 77°F (25°C) should be between 2.2 and 2.6 ohms.

c.   Installation.

(1) Install rotor bearing, refer to paragraph 5-8.

(2) Install rotating rectifier on rotor, refer to paragraph 5-5.

(3) Install generator, refer to paragraph 2-10.

5-5.   **ROTATING RECTIFIERS (DIODES).**

<u>CAUTION</u>

Do not attempt shaft.   remove exciter rotor from main rotor the windings will result
Damage is attempted.   if removal
              to
              to

a. <u>Testing.</u> A shorted or open diode can cause poor generator operation.This is evident by failure of the generator terminal voltage to build up to rated value or a terminal voltage that is too low.

MEP-021B (400 Hz)   MEP-016B/MEP-026B (60 Hz & 28 VDC)

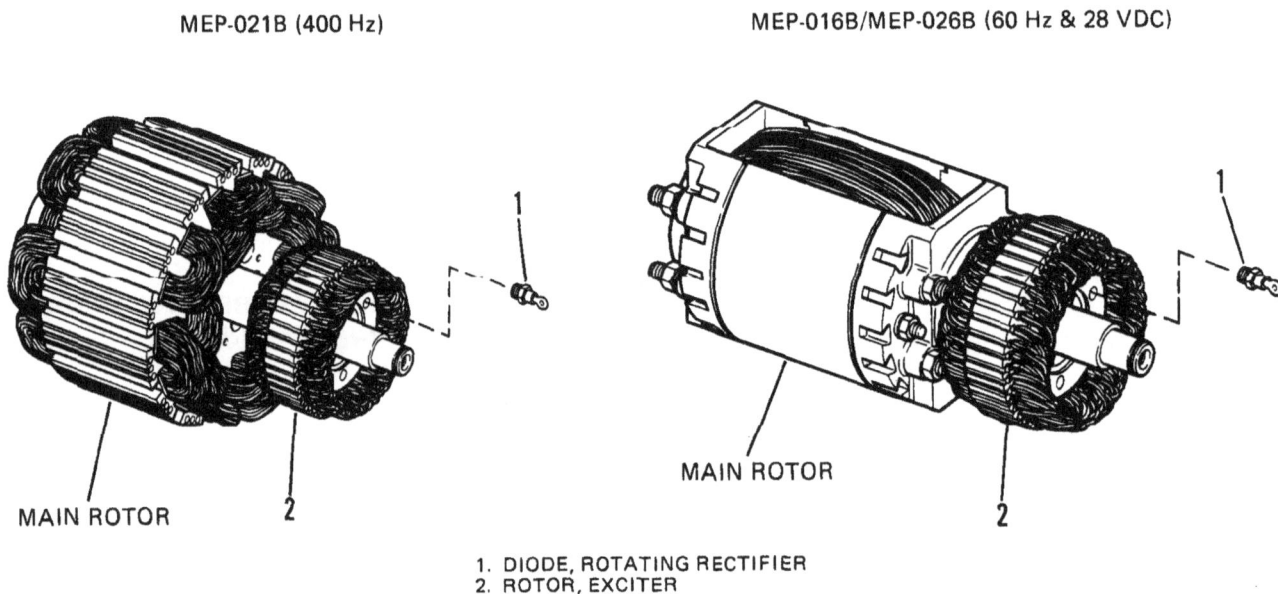

MAIN ROTOR   2   MAIN ROTOR   2

1. DIODE, ROTATING RECTIFIER
2. ROTOR, EXCITER

(1) Remove diodes (1, Figure 5-9) from exciter rotor, refer to b, Removal.
Figure 5-9. Rotating Rectifiers.

(2) Use a multimeter (set to read ohms) to measure the resistance between the terminal end and the threaded base (heat sink) end of the diode. Reverse
multimeter leads and repeat the resistance measurement.A diode in good condition will have a very high resistance for one measurement and a resistance near zero when multimeter probes are reversed.Failure to obtain these two extremes in resistance measurement indicates a defective diode that should be replaced.

NOTE

Be sure points of multimeter probes are sufficiently long and sharp to penetrate any insulating varnish on diode terminals.

b. Removal.

(1) Remove rotor, refer to paragraph 5-4.

CAUTION

To protect diode from damage, hold diode with needle nose pliers or other suitable heat sink.   Use a low melting point solder. Diode can be damaged if allowed to overheat during soldering.

(2) Hold diode (1) lead with a heat sink (needle nose pliers, alligator clip, etc.) to prevent the transfer of heat to the diode as it is being unsoldered.

(3) Unsolder lead   wire from diode with a 25 to 40 watt soldering iron, and
    remove diodes   (1) from exciter rotor (2).

c.   Installation.

(1)   Install diodes (1) into exciter rotor, tighten diodes to 12 to 15 in lbs
       (0.27 to 0.34 N.m).

(2)   Using a 25 to 40 watt soldering iron, solder lead wires to terminal of
       diode.   To protect diode from damage due to heat caused by soldering, hold
       diode terminal with needle nose pliers as a heat sink and use a low melting point solder.

       Install rotor, refer to paragraph 5-4.
(3)

WARNING

Observe safety regulations. The voltages used in this test are dangerous to human life. Contact with the leads or the windings under test may cause severe and, possibly, fatal shock. Arrange the high voltage leads so that they are not in a position to be accidentally touched. Keep clear of all energized parts. Always reduce the test voltage to zero and ground the winding under test before making any mechanical or electrical adjustments on the equipment. When grounding out windings which have been tested, always connect the connection wire to ground first, and then to the winding. Never perform other person assisting. securely grounded.

this test without at least one Generator frame

a.   Perform high potential test on   stator.   Do not perform high potential test of
      exciter stator.

(1)   General.The high potential test is performed to determine whether or not the insulation of the equipment under test
is defective.   It is customary to determine whether electrical equipment will withstand normal voltage stresses by means of a test in which higher voltage
than normal are applied for a definite period of time. The applied voltage must not be so high
(above 1,500V) as to damage the insulation unless the insulation was initially defective.

Test Equipment.Use acceptable high potential test equipment, following the operating instructions of the manufacturer. Refer to Table 2-1 or
Appendix B, Section III of TM 05926B/06509B-12/TM 5-6115-615-12/NAVFAC P-8-646-12/TO 35C2-3-386-31
(2)   manual.

Procedure.
(a) Adequately ground high potential test equipment (Figure 5-10) to a water

(3)

pipe or similar electrical ground accompanying the

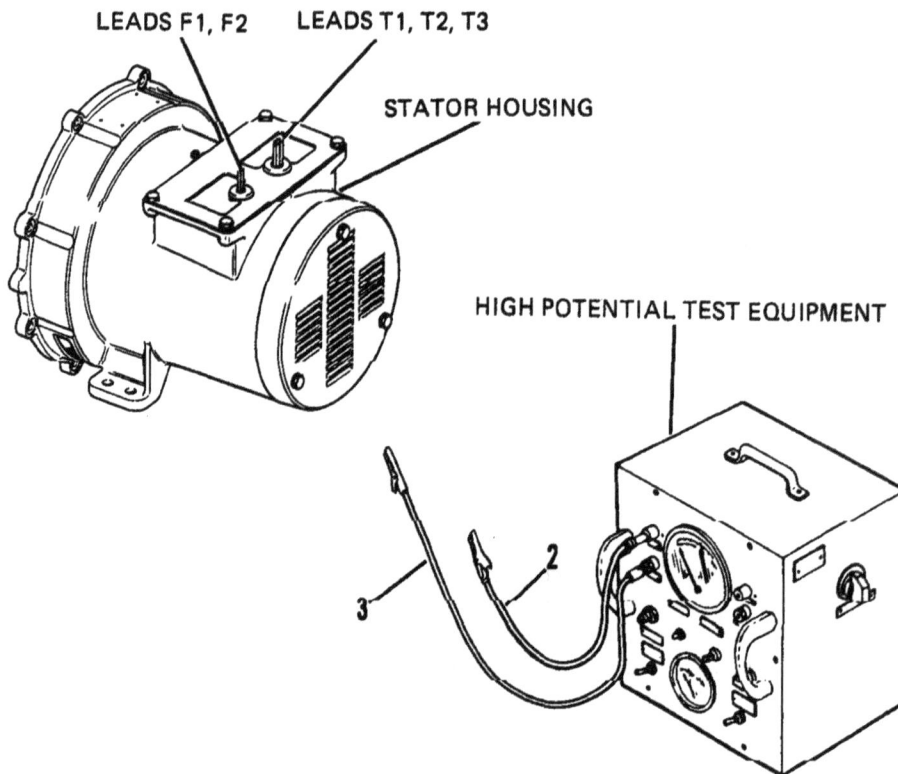

Figure 5-10. Stator Winding Insulation Test.

(b) Connect the high-voltage lead (2) from the test   equipment to the winding
     (T1, T2, or T3) being tested.

NOTE

On stator for sets MEP-016Band MEP-021B (60 and 400Hz) sets there are six stator leads (T1, T2, T3, T4, T5, and T6). The stator for the MEP-026B generator set has three leads (T1, T2, and T3).   The insulation test connections will only involve leads (T1, T2, and T3) all three sets.

on

(c) Connect the ground lead (3) (or place stator on grounding bed) to the stator frame.Make sure ground lead has a good, clean connection.

(4) Test.

(a) Turn on test equipment in accordance with manufacturer's instructions after making sure that the initially applied voltage will be not greater than 600 volts.

(b) The test voltage then shall be raised uniformly to the required maximum (1,500 Volts RMS, 60 Hz). This increase shall be accomplished in not less than 10 seconds nor more than 30 seconds.

(c) Apply the maximum voltage for 1 minute.

(d) After 1 minute of applied maximum voltage, the voltage then shall be reduced gradually to the voltage initially applied.   This reduction shall not be accomplished in less than 5 seconds.

(e) Turn off test equipment. Ground the high voltage lead of the test equipment to make sure that no charge remains on

CAUTION
- - - - - - - - - -

the windings which have been under test.

(f) Remove the high-voltage lead (2) from the winding and proceed with the tests of the remaining circuits to be tested.Make sure that all circuits not under test are securely grounded.

(5) Results.Any evidence of insulation breakdown is cause for replacement of the stator under tests.

h   Perform winding resistance test.   Refer to Figure 5-11.

(1) Perform winding resistance test using a multimeter (set to read ohms).

(a) On MEP-026B (28 VDC) set measure and T1-T3. resistance between leads T1-T2, T2-T3,

(b) On MEP-016B/MEP-021B (60/400 Hz)   sets measure between leads T3-T6, T1-T4, and T5-T2.

MEP-016B/MEP-021B
(60/400 Hz)

MEP-026B (28 VDC)

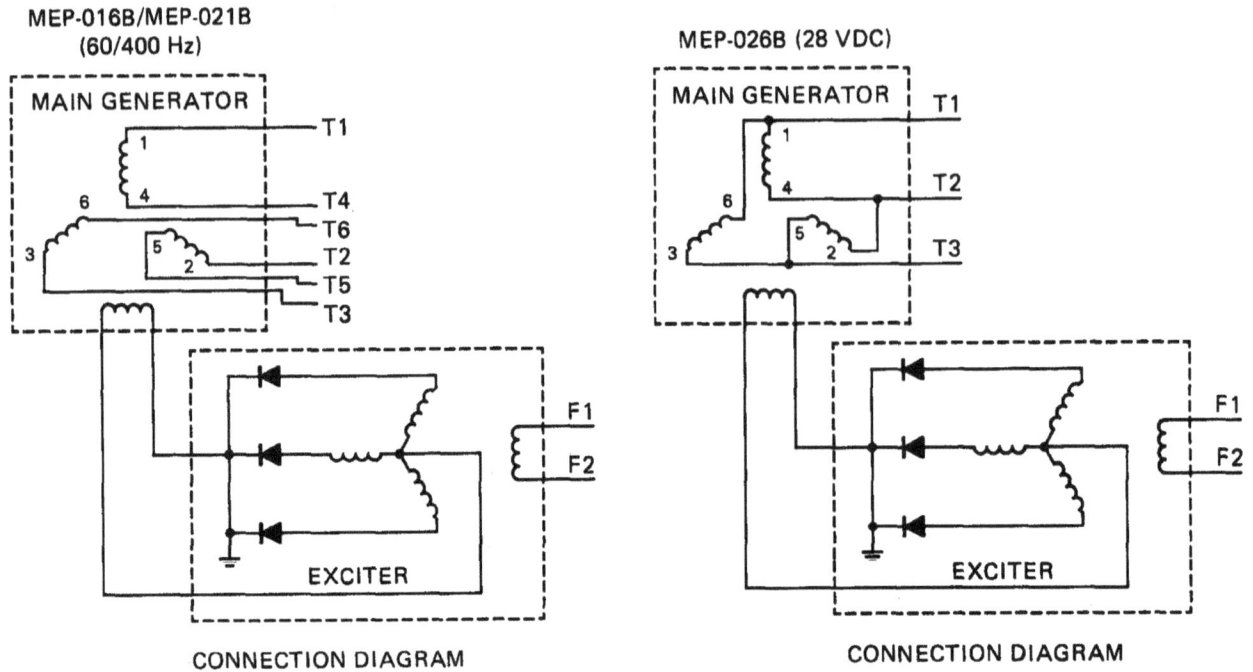

Figure 5-11.Generator Wiring Schematics.

(2) Winding resistance should be as follows:

  (a) MEP-016B set - the main stator winding resistance between lead pairs and 0.370 ohms.
should be between O.304

  (b) MEP-021B set - the main stator winding resistance between lead pairs and 0.181 ohms.
should be between 0.149 stator winding resistance between leads should be ohms.

  (c) MEP-026B set - the main
between 0.018 and 0.022

(3) A winding resistance above or below the resistance values given indicates a defective stator and stator should be replaced.

5-7. EXCITER STATOR TESTING.

NOTE

Testing, removal, and installation of the exciter rotor is described in paragraph 5-4.The exciter rotor is not removed from the rotor shaft.The following procedure covers the exciter stator only.

CAUTION

Do not perform high potential insulation test on exciter stator.

a. Separate the stator leads F1 and F2   from the other leads.

b. Use a multimeter (set to read ohms)   to measure the resistance of the
    windings.

c. Exciter stator windings should have   a resistance between 20.34 and 24.86
    ohms.

5-8. GENERATOR   BEARING.

a. Removal

(1) Remove   rotor, refer to paragraph 5-4.

(2) Remove   circlip (1, Figure 5-12).

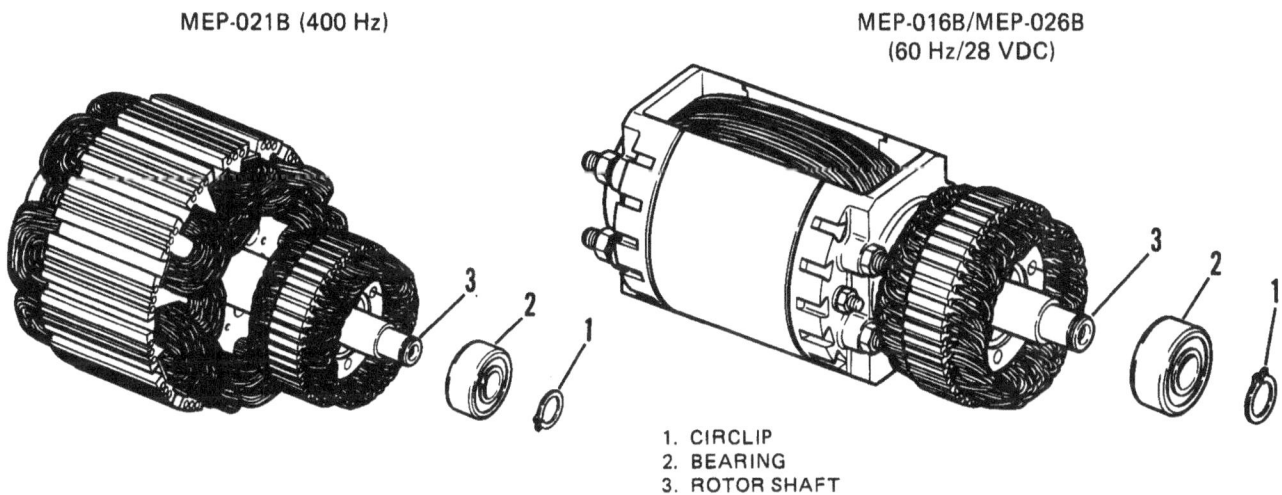

MEP-021B (400 Hz)   MEP-016B/MEP-026B
                           (60 Hz/28 VDC)

1. CIRCLIP
2. BEARING
3. ROTOR SHAFT

Figure 5-12. Generator Bearing.

(3) Remove bearing (2) from rotor shaft (3) with a puller.

b.   Installation

(1) Apply a light coating of grease to bearing.

CAUTION

Press bearing on inner race only.

(2) Press bearing (2) on to rotor shaft (3) until bearing contacts shoulder.

(3) Install circlip (1).

Install rotor, refer to paragraph 5-4.
5-9.VOLTAGE REGULATOR.

a. Testing.

(1) Remove voltage regulator, refer to step c, Removal.

(2) Test power transistor Q2.

regulator housing.

FRONT VIEW

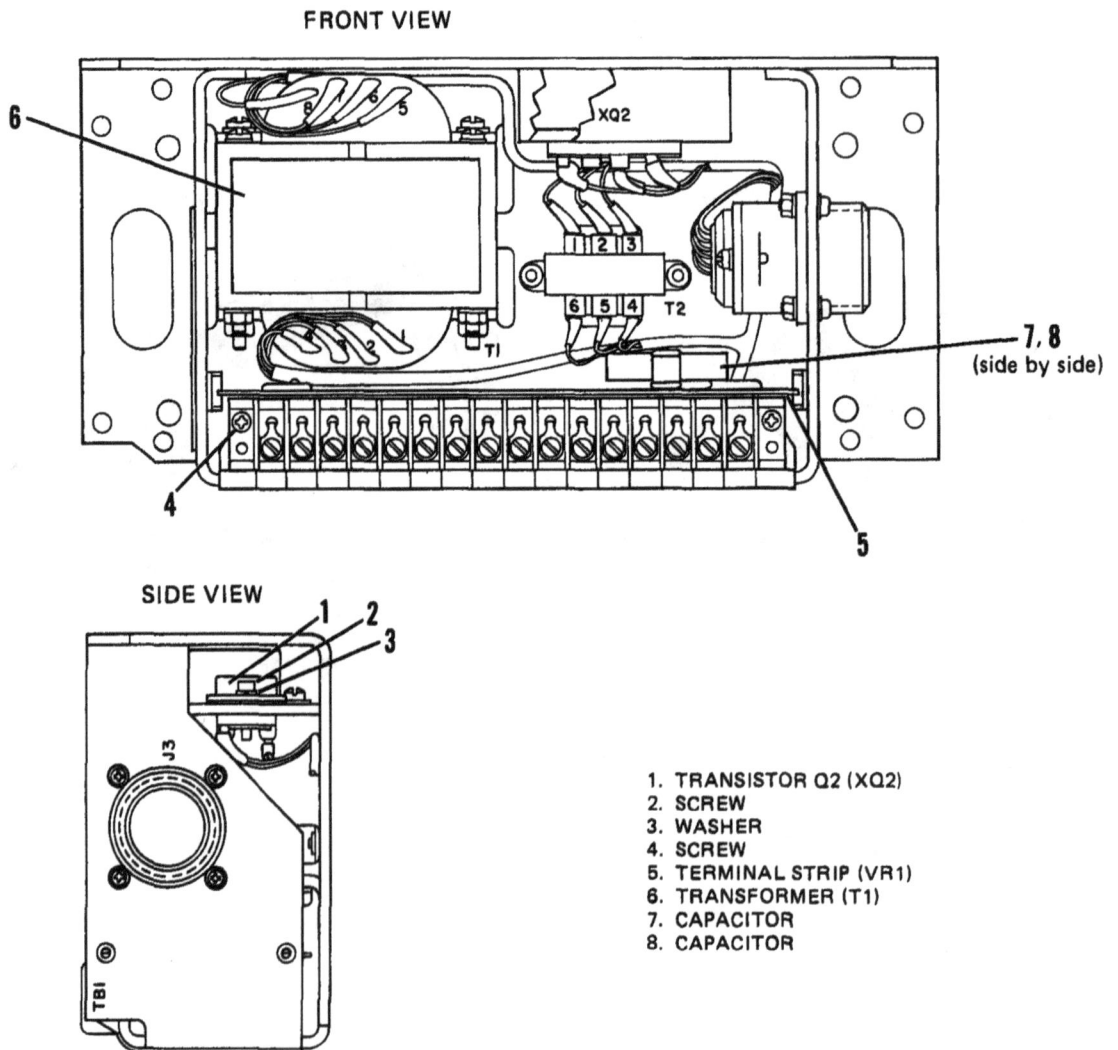

SIDE VIEW

1. TRANSISTOR Q2 (XQ2)
2. SCREW
3. WASHER
4. SCREW
5. TERMINAL STRIP (VR1)
6. TRANSFORMER (T1)
7. CAPACITOR
8. CAPACITOR

Figure 5-13. Voltage Regulator Testing.

(b)   Remove screws (2) and washers (3) securing Q2 to housing.

(c)   Pull transistor (1) straight from housing, being careful not to damage themica insulator under the transistor. Do not remove the white heat sink paste from the mica insulator, transistor, or regulator

CAUTION

housing.   If this compound is removed transistor Q2 will overheat and become damaged.

Test transistor with a multimeter (set to read ohms).   Refer to Figure 5-14
and Table 5-1. Multimeter (+) and (-) refer to the multimeter test leads.

(d)   **Figure 5-14. Q1and Q2Transistor Pin Out Locations.**

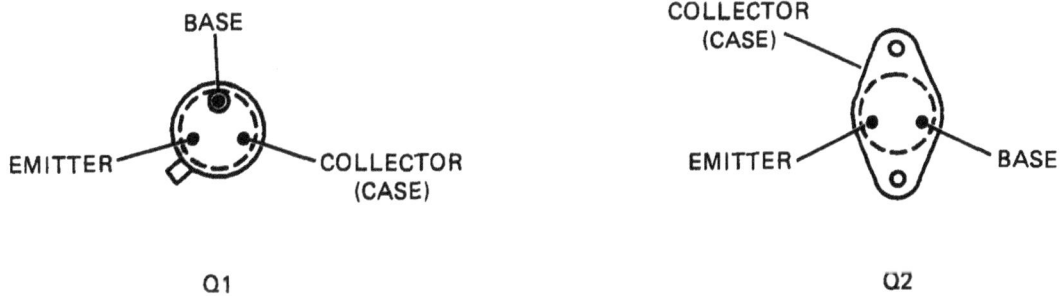

Q1

Q2

Table 5-1. Q2Test Chart.

Multimeter (+)	Multimeter (-)	Meter Reading	
Collector	Emitter	Infinite	(∞)
Collector	Base	Infinite	(∞)
Emitter	Collector	Infinite	(∞)
Base	Collector 1 to 50ohms		
Base	Emitter	1 to 50 ohms	
Emitter	Base	Infinite	(∞)

(e) Readings other than those listed in Table 5-1 indicate either an open
or shorted junction within Q2 and Q2 should be replaced.

NOTE

Do not install transistor Q2 at this time. With Q2 removed, transistor Q1 can be
tested without removal of the transistor from the printed circuit board.

(3) Remove terminal strip   VR1.

    (a) Remove two screws   (4, Figure 5-13   (4) Test diode CR5.
                                                     that secure terminal strip VR1 (5).

    (b) Gently pull terminal strip VR1 (5)
        upward.

    (a) Locate diode CR5 located on circuit board.

    (b) Test diode CR5 with a multimeter (set to read ohms). Measure resistance across CR5, reverse test
        leads, and measure resistance again.

    (c) Resistance should be low (1 to 50 ohms) one way and infinite

                                                                               (∞)

        resistance the other way.

    (d) If low or high resistance is measured in both directions diode CR5
        should be replaced.

(5) Test current regulating diode CR6.

    (a) Locate diode CR6 located on circuit board.

    (b) Unsolder and lift one end of diode CR6 from the printed circuit board.

    (c) Test diode CR6 with a multimeter (set to read ohms). Measure resistance across CR6, reverse test leads,
        and measure resistance again.

    (d) Resistance should be low (1 to 50 ohms) one way and greater than 500
        ohms resistance the other way.

    (e) If low or high resistance is measured in both directions diode CR6
        should be replaced.

NOTE

Do not install diode CR6 lead to printed circuit board until transistor Q1 has been tested.

(6) Test diode CR7.

    (a) Locate diode CR7 on circuit board.

(b) Test diode CR7 with a multimeter (set to read ohms). Measure resistance across CR7, reverse test leads, and measure resistance again.

(c) Resistance should be low (1 to 50 ohms) one way and infinite resistance the other way. **(∞)**

(d) If low or high resistance is measured in both directions, diode CR7 should be replaced.
(7) Test transistor Q1.

(a) Make sure that transistor Q2 is still removed and that one lead of diode CR6 is unattached to the printed circuit board.

(b) Locate power transistor Q1 on the printed circuit board.

(c) Test transistor with a multimeter (set to read ohms). Refer to Figure 5-14 and Table 5-2. Multimeter (+) and (-) refer to the multimeter test leads.

Table 5-2. Q1 Test Chart.

Multimeter (+)	Multimeter (-)	Meter Reading
Collector	Emitter	Infinite **(∞)**
Collector	Base	Infinite **(∞)**
Emitter	Collector	Infinite **(∞)**
Base	Collector 1 to 50 Ohms	
Base	Emitter	1 to 50 Ohms
Emitter	Base	Infinite **(∞)**

(d) Resistance measurements other than those listed in Table 5-2 indicate an open or shorted junction within Q1, Q1 should be replaced.

(8) Solder lead of diode CR6 to printed circuit board (6, Figure 5-13).

(9) Install transistor Q2 (1, Figure 5-13) with mica insulator between Q2 and regulator housing.

(10) If necessary, apply a small amount of heat sink compound (Table 2-2, item 14) (P/N 13217E3704, FSCM 97403) to transistor Q2 and housing. Secure transistor Q2 with screws (2) and washers (3).

(11) Make test connections.

(a) Connect a 30 ohm 120 watt resistor to terminals #21 and #20 of terminal strip VR1 (5).

(b) Obtain a 750 ohm 2 watt rheostat.

MARINE CORPS TM 05926B/06509B-34
ARMY   TM 5-6115-615-34
NAVY   NAVFAC P-8-646-34

(c)   Position rheostat so that control shaft is facing you and the three terminal lugs are facing up.Connect a 12-18 AWG insulated wire to the middle terminal lug and another wire to the left hand terminal lug of the rheostat.

(d)   Attach the two rheostat leads to terminals #26 and #28 of terminal strip VR1 (5).Turn the rheostat clockwise to its maximum rotation.

(e)   Connect a 0-150 VAC, 60 HZ power supply (refer to Table 2-1) to terminals #24 and #27 of terminal strip VR1 (5).

(12) Perform tests.

(a)   Turn on the power supply and set for 115 VAC.

(b)   Set a multimeter to read a 0-50 VDC range.

(c)   Check for 30-35 VDC between terminals #20 and #21 of terminal strip VR1 (5).Terminal #20 is positive and terminal #21 is negative.

(d)   Rotate the shaft of the test rheostat counterclockwise to approximately eighty percent of its maximum rotation. The voltage between terminals #20 and #21 of VR1 (5) should now be zero.

(e)   Rotate the shaft of the test rheostat clockwise while observing the voltage at terminals #20 and #21 of VR1 (5).

(f)   Voltage at terminals #20 and #21 should be between 30 and 35 volts when test rheostat is turned to approximately eighty percent of its clockwise rotation.If no voltage is present, refer to step (i).
If there is a delay of two seconds or more between decreases and increases of the voltage at terminals #20 and #21 or if voltage changes do not occur at the eighty percent of maximum rotation points, adjust R6.   Refer

(g)   to (h) for adjustment procedure.   If no delay is present and voltage changes occur at the eighty percent rotation points, testing procedure is completed.   If no voltage change occurs, regardless of the position of the test rheostat, refer to step (i).

(h)   Locate R6 on the printed circuit board (6). Turning the adjustment screw of R6 clockwise will turn the voltage at terminals#20 and #21 of VR1 (5) on for a given setting of the test rheostat.   Turning adjustment screw counterclockwise turns the voltage at terminals #20 and #21 off. Adjust screw as required to achieve voltage cutoff and turn on for eighty percent of maximum rotation as described in steps (d), (e), and (f).   Troubleshooting is complete after adjustment is made.

NOTE

Adjustment of R6 does not cause the voltage at terminals #20 and #21 of VR1 (5) to vary. Either voltage will be on or voltage will be off.

WARNING

Turn off 115 VAC power source before attempting any inspection or repair.

(i)   Inspect circuit board for burned or broken printed circuit board
paths.   If burned or broken paths are present, circuit board should be replaced.

(j)   Locate components R1 and R2. Refer to Figure 5-15 for a wiring schematic. Apply 115 VAC to terminals
#24 and #27 of terminal strip VR1 (5, Figure 5-13).

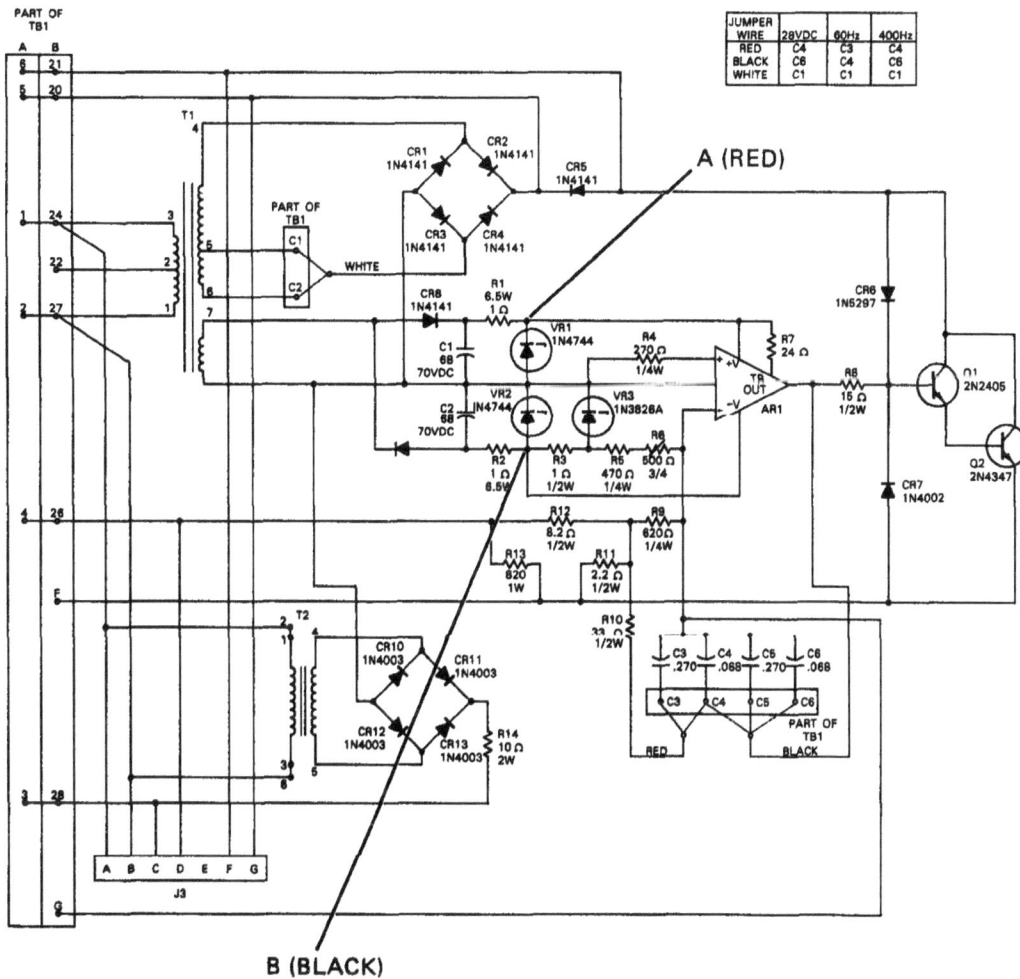

Figure 5-15. Voltage Regulator Schematic.

5-23

(k)   Connect a multimeter (set to read 0-50 VDC) across points "A" and "B" as indicated on wiring schematic. Connect red lead to point "A" and black lead to point.
A DC voltage of 28-32 volts should be indicated.   If voltage is not within this

(1)   range, check the voltage diode VR1 and VR2 on circuit board.
If either zener diode VR1 and VR2 is removed, do not apply the 115 VAC source to across voltage regulating (zener)

CAUTION

the regulator until a replacement is connected into the circuit.Failure of the voltage regulator will occur if this caution is not observed.

The voltage across VR1 and VR2 should be approximately 14 VDC.   Replace zener diodes that do not have a voltage drop of approximately 14 VDC.

(m)   After replacing zener diodes VR1 and/or VR2 and no voltage is indicated across either or both zener diodes VR1 and VR2, check the voltage across terminal lugs #7 and #8 of the T1 power transformer (6).

(n)

NOTE

Change setting on   multimeter to 0-50   VAC range.

(o)   A voltage of 25-35 VAC   should be present.   If voltage is not present,
check that 115 VAC power source is connected properly and all wiring connections to T1 (6) are tight.   If power is properly connected and T1
(6) connections are tight transformer T1 (6) should be replaced.

(p)   After verifying that 25-35 VAC is present across T1 (6) terminals #7 and #8, locate diodes CR8 and CR9 on circuit board (6).

(q)   Turn off power and unsolder one end of diode CR8 and CR9. Check the resistance of each diode in one direction and then reverse leads and test in the other direction. and high in the other.
Resistance should be low in one direction

(r)   Replace diodes CR8 or CR9 if   they do not pass resistance check.

(s)   Locate capacitors C1 (7) and circuit board.   C2 (8) on the reverse side of the printed end of each capacitor Unsolder one board.   from the circuit

(t)   Test capacitors C1 (7) and C2 (8) with a multimeter (set to read ohms) across the capacitor being tested and then reverse leads across

(u) Connection of the multimeter (set for ohms) across the capacitor should immediately result in the movement of the meter pointer from the infinity   side of the scale to the low ohms side of the scale and then slowly return to the infinity (∞)de of the scale.
If the capacitor is shorted or leaky, the pointer will remain on the low ohms side of the scale.   If the capaci-

(v) tor is open, there will be no movement of the meter pointer.   Replace open, shorted, or leaky capacitors as required.

(w) Locate zener diode VR3 on circuit board and apply 115 VAC to terminals #24 and #27 of the terminal strip VR1(5, Figure 5-13).

(x) Check the voltage across zener diode VR3.Voltage across VR3 should be approximately 4.7 VDC.

(y) If voltage across VR3 is greater than 4.7 VDC, replace VR3.If no voltage is indicated across VR3, locate R3 a 1K ohm resistor located on circuit board.

(z) Disconnect the 115 VAC power source from the regulator, unsolder one end of R3 from the circuit board. Measure the resistance of R3, resistance should be between 900 and 1100 ohms.   If resistor is within this range, reconnect lead.Replace resistor if not within tolerances.

(aa) Apply the 115 VAC power to terminals #24 and #27 of the terminal strip VR1 (5). Recheck the voltage drop across zener diode VR2, there should be a voltage drop of 14 VDC present. If 14 VDC is present and there is no voltage drop across zener diode VR3, VR3 should be replaced.

(bb) If none of the tests or recommended repairs correct the operation of the voltage regulator, entire voltage regulator should be replaced.

b.Adjustment.

(1) The following components and equipment are needed for the adjustment procedures on the voltage

- DC Voltmeter ± 0.5 percent,   0-100 VDC (M1 and M4) 0-150
- AC Voltmeter ± 0.5 percent,   VAC (M2)
- Oscilloscope (M3)
- Voltage source 0-150 VAC 60

  ±0.25 percent, from 0-2 Amps (PS1) Resistor 30 ohms,
- 120 watts (R1)
- Resistor, 750 ohms ± 5 percent 2 watts (R2)

(2) Connect voltage regulator, equipment, and components as shown in Figure 5-16.

(3) Set potentiometer R6 to 250 ohms (halfway).

(4) Turn AC power supply (PS1) on.

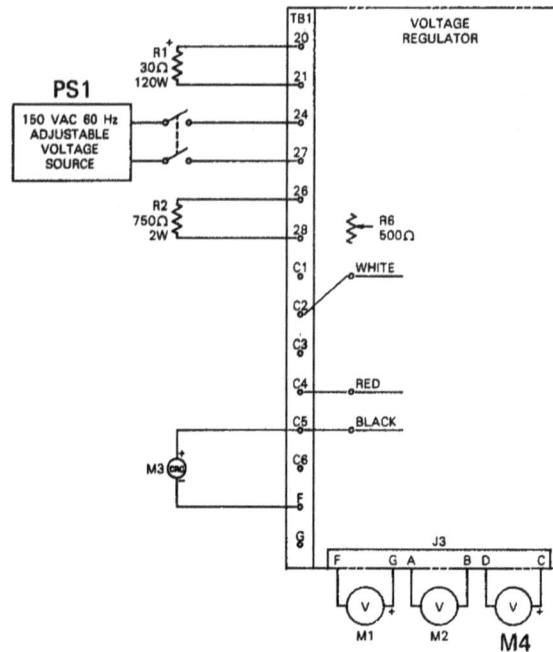

Figure 5-16. Voltage Regulator Adjustment Schematic.

(5)   Slowly increase AC voltage (0 to 30 VAC) while monitoring DC voltmeter (M1) and oscilloscope (M3).DC voltage reading on DC voltmeter (M1) shall increase proportionally with the increasing AC voltage and the voltage wave form trace of oscilloscope (M3) shall move toward a positive voltage level. A voltage shall also be indicated on meter DC voltmeter (M4).The waveform trace of oscilloscope (M3) swinging below the zero axis (negative) during this (0-30 VAC) test shall be indicative of an unusable/defective operational amplifier.

Increase AC voltage for a "maximum reading" on meter (M1) (35 VDC minimum) or until oscilloscope (M3) wave form is just starting to come off the "maximum level" refer to curve 1 of Figure 5-17.

(6)

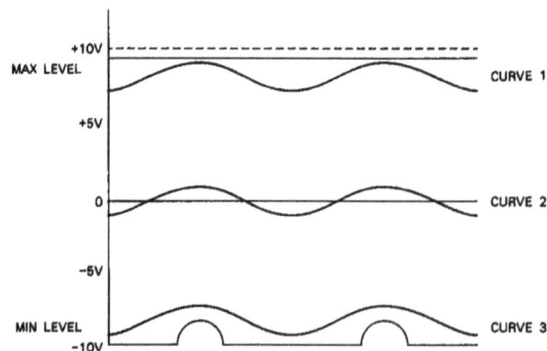

Figure 5-17. Waveforms.

    (a) If the AC voltmeter (M2) indicates 136 VAC or greater, proceed with
        step (7).

    (b) If the AC voltmeter (M2) indicates less than 136 VAC, proceed with step
        (8).

(7) Perform the following adjustment procedures:

    (a) Adjust AC supply (M2) to 136 VAC.

    (b) Slowly adjust R6 clockwise until oscilloscope (M3) wave form begins to decrease from "maximum level".
DC voltmeter (M1) reading shall go to zero when wave form passes through zero.   Refer to curve 2 on Figure
5-17.

    Adjust AC input voltage AC voltmeter (M2) to 133 VAC.The oscilloscope
        (M3) waveform shall return to maximum level and DC voltmeter (M1) shall read maximum (35
    (c)

    VDC minimum).

(8) Perform the following adjustment procedures.
Adjust AC supply (PS1) to adjust (M2) to 133 VAC.

    (a) Slowly adjust R6 counterclockwise until oscilloscope (M3) wave form begins to go up from "mini-
mum level" (curve 3, Figure 5-17). Ml reading shall go to "maximum level"(35 VDC minimum) when
    (b) wave form passes through zero.

    Adjust AC supply (PS1) to adjust (M2) to 136 WAC.The oscilloscope
        (M3) wave form shall return to "minimum level"and Ml shall read zero volts.

    Without further adjustment of R6, repeat the above procedure for the remaining three conditions tabu-
    (c)
lated in Table 5-3.In all cases, similar performance should be achieved.

NOTE
(9)

    Test condition one in Table 5-3 has already been accomplished in the previous pro-
cedures.Procedures are repeated for the remaining three conditions.

Table 5-3.Test Conditions.

Condition	White Lead	Black Lead	Red Lead	M3+ Lead
1	C2	C5	C4	C5
2	C2	C5	C3	C3
3	C1	C4	C3	C4
4	C1	C6	C4	C6

NOTE

Failure of indicative the unit to meet adjustment requirements is of a defective regulator.

(10) After all adjustments have been made, item 13) between the adjustment screw screw head. apply a locking compound (Table 2-2, head and housing surfaces adjacent to

c. Removal.

NOTE

Both the MEP-016B/MEP-021B (60/400) Hz sets and the MEP-026B (28 VDC) set have similar procedures for removing and installing the voltage regulator. Refer to Figure 5-18 for MEP-016B/MEP-021B (60/400 Hz) sets and Figure 5-19 for MEP-026B (28 VDC) sets.

(1) Tag and disconnect wiring from voltage regulator (3, Figure 5-18 or 5-19).

(2) Remove nuts (1), screws (2) and regulator (3).

d. Cleaning, Inspection, and Repair.

(1) Clean voltage regulator with compressed air and a soft bristle brush.

(2) Inspect voltage regulator for cracks, damaged terminals, or other visible damage.

(3) For testing, refer to step a, Testing.

(4) Repair of components identified by testing as faulty is by component replacement.

CAUTION

Solder joints and/or component leads (except potentiometer leads) should not project more than 0.125 in. beyond the surface of the board.

(a) Transistor replacement.

1 Make sure that replacement transistor has same identification number as the original transistor.

2 Apply heat sinks to the circuit board runs connected to the transistor being replaced.

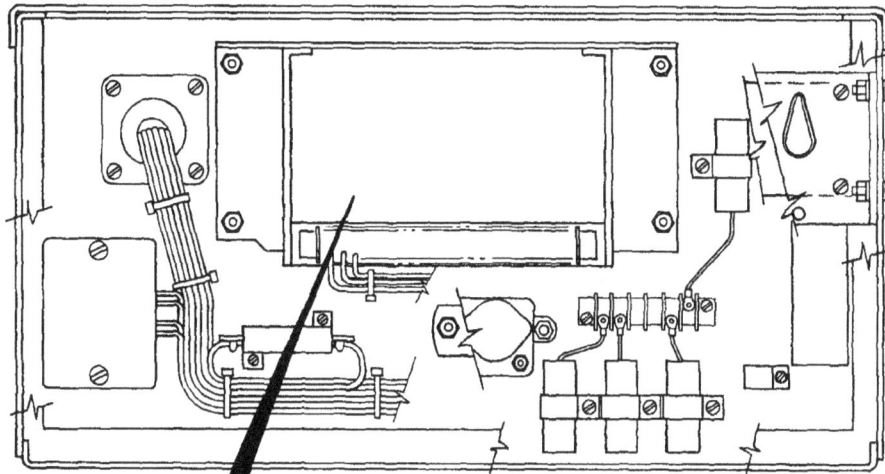

MEP-016B SHOWN, MEP-021B SIMILAR

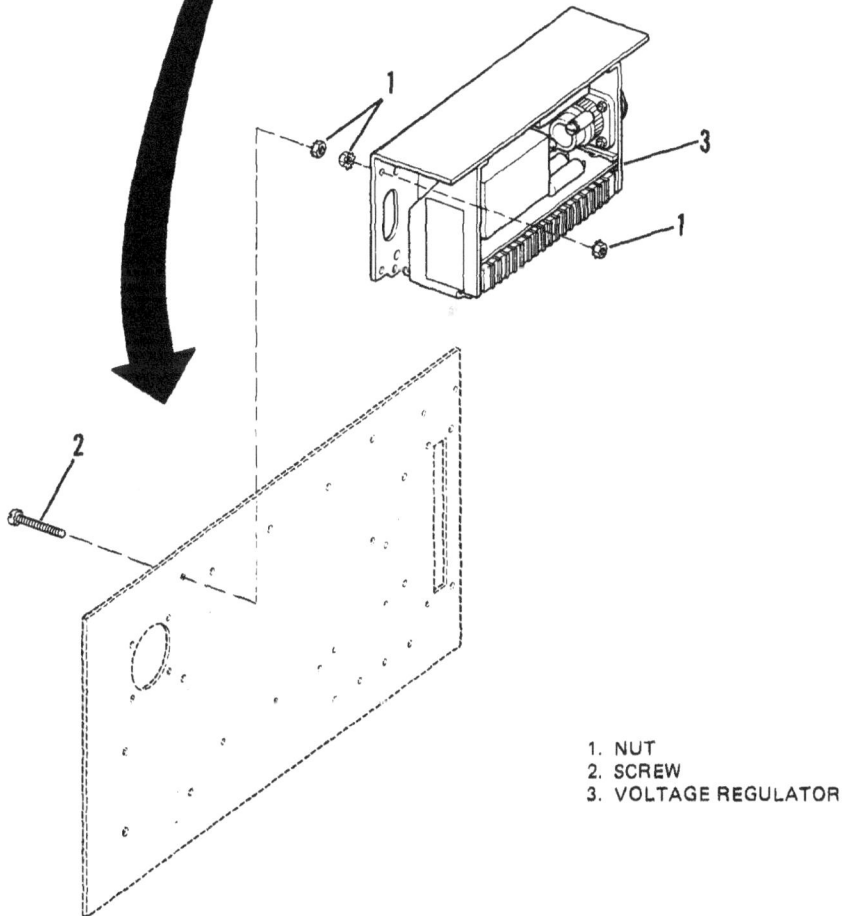

1. NUT
2. SCREW
3. VOLTAGE REGULATOR

Figure 5-18. Voltage Regulator. MEP-016B and MEP-021B (60/400 Hz Sets).

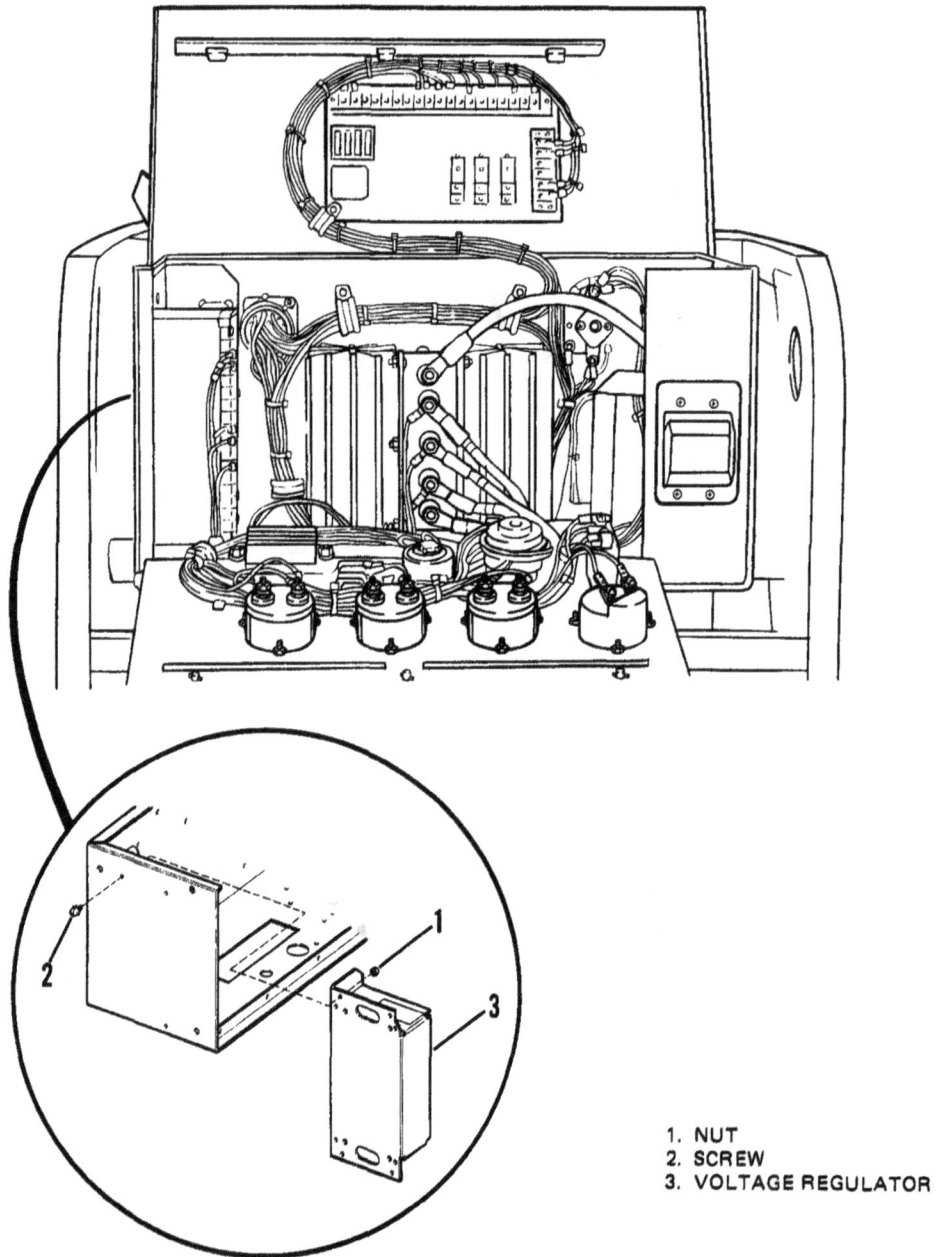

1. NUT
2. SCREW
3. VOLTAGE REGULATOR

Figure 5-19.   Voltage Regulator MEP-026B (28VDC Set).

3 ___ Note and record position of transistor leads and tab on transistor case in relation to the printed circuit board.

4 ___ Unsolder transistor from circuit board and remove.

Make sure emmiter, collector, and base leads are positioned as noted during removal.

6 ___ Solder transistor in accordance with MIL-STD-454 Requirement 5. Trim transistor leads.

e.   Installation.

(1) Check that red, black, the generator   and white jumper wires are properly positioned for Refer to Figure 5-20.
     output.

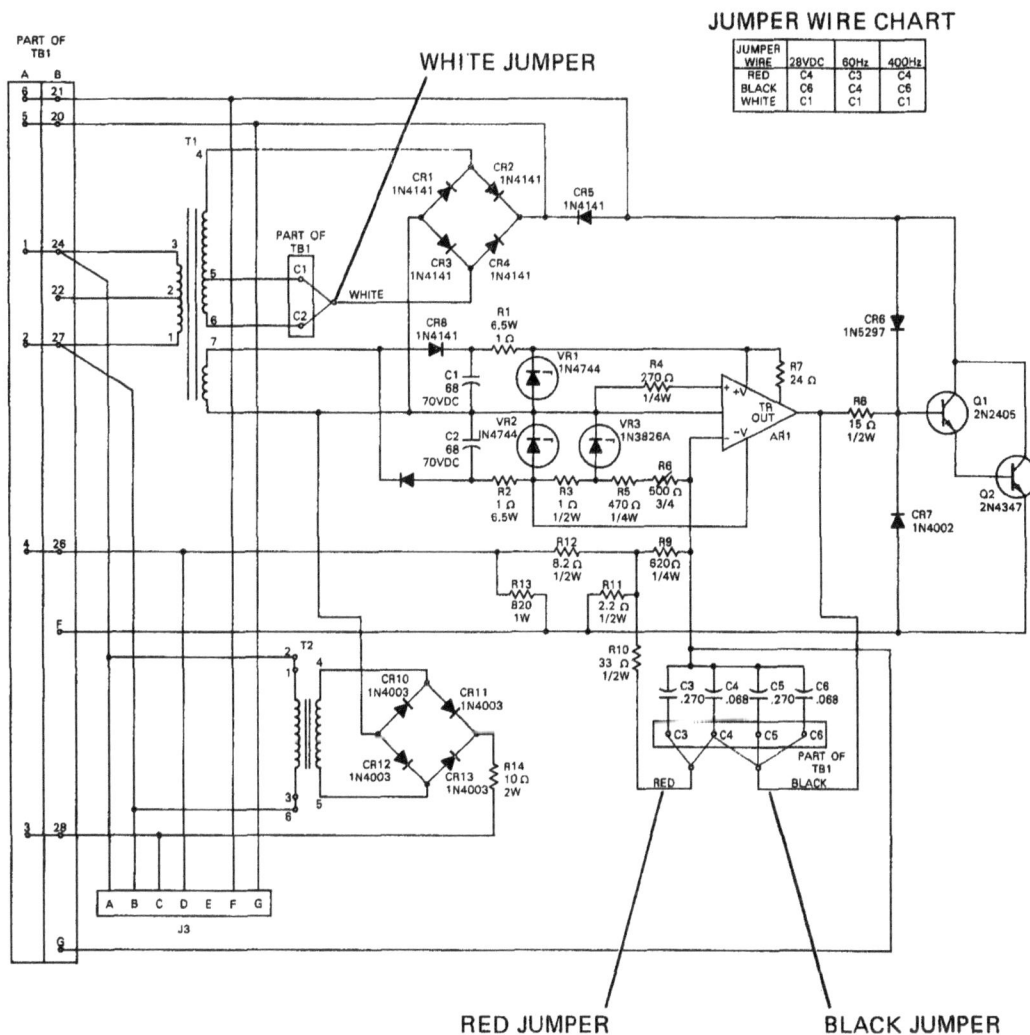

JUMPER WIRE CHART

JUMPER WIRE	28VDC	60Hz	400Hz
RED	C4	C3	C4
BLACK	C6	C4	C6
WHITE	C1	C1	C1

WHITE JUMPER

RED JUMPER   BLACK JUMPER

Jumper Wire Placement.

(a)   MEP-026B (28 VDC)-Red jumper wire is connected to terminal C4,Black jumper wire is connected to terminal C6, and the White jumper wire is connected to terminal C1.

(b)   MEP-021B (400 Hz)-Red jumper is connected to terminal C4, Black jumper wire is connected to terminal C6, and white jumper wire is connected to terminal C1.

(c)   MEP-016B (60 HZ)-Red jumper wire is connected to terminal C3,Black jumper wire is connected to C4,and white jumper wire is connected to terminal C1.

Place voltage regulator (3, Figure 5-18 or 5-19) in control box and secure with screws (2) and nuts (1).

Use tags for identification to connect leads to voltage regulator.

(3)   **5-10.   CURRENT TRANSFORMER MEP-016B/MEP-021B (60/400 Hz) SETS ONLY.**

a.Testing

(1)Tag and disconnect wires and resistors from transformer (3, Figure5-21)

terminals.

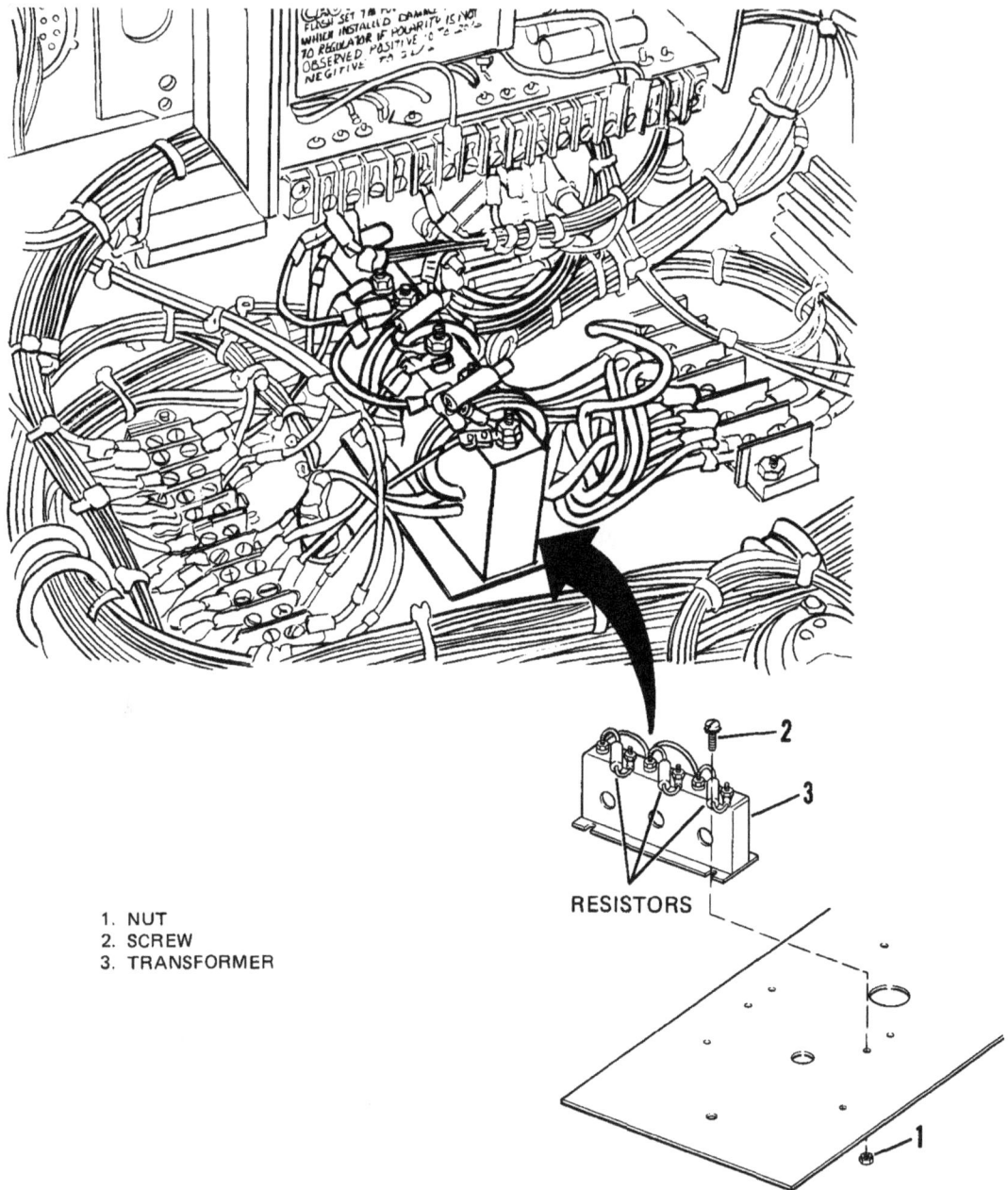

1. NUT
2. SCREW
3. TRANSFORMER

Figure 5-21.Current Transformer.

(2)   Use a multimeter (set to read ohms) to check the transformer windings.
Check resistance across terminals A-1 and A-2, B-1 and B-2, C-1 and C-2.

(3)   Multimeter should indicate an equal resistance between each terminal pair.
Resistance should be between 12 and 18 ohms.

b.   Removal.

(1)   Tag and disconnect wires and resistors from transformer (3, Figure 5-21) terminals.

Tag and disconnect leads that pass through current transformer from the circuit breaker.
(2)

Remove nuts (1), screws (2), and transformer (3).

(3)

c.   Installation.

(1)   Secure transformer (3) with screws (2) and nuts (1).

(2)   Thread leads through transformer and attach to circuit breaker.   Use tags as
identification.

(3)   Connect wires and resistors to transformer using tags as lead identification.

CHAPTER 6

MAINTENANCE OF THE FUEL SYSTEM

6-1.GENERAL. Theengine fuel system consists of two pressure systems (a low pressure system and a high pressure system). In the low pressure system fuel is pumped from the main fuel tank through a fuel/water separator to the injection pump by a mechanical fuel transfer pump mounted on the engine.   If an auxiliary fuel source is selected, fuel is pumped by an electric fuel pump from the auxiliary fuel source to the main fuel tank.The electric fuel pump is controlled by a fuel high/low level switch on the main fuel tank.The high pressure fuel system consists of the fuel injection pump, injection line, and fuel injector.

6-2.FUEL TANK.

a.<u>Removal.</u>

(1) Drain the fuel tank.

(2) Remove frame, refer to paragraph 3-2.

(3) Remove nut (1, Figure 6-1), lockwashers (2), and washer (3).

(4) Remove strap (4), gasket (5).

(5) Remove screw (17) that secures bracket (18) to frame. Remove tank (6). b.   <u>Disassembly.</u>

(1) Remove filler cap (7) and fittings (8, 9 and 10).

(2) Remove drain valve (11) and fitting (12).

(3) Remove level switch (13).

(4) If necessary, remove nut (14), screw (15)   and valve cap and retainer
assembly (16).

(5) Remove screws (19), flange (20) and gasket (21). c.<u>Cleaning Inspection and</u>
<u>Repair.</u>

(1) Wipe all unpainted metal parts with a clean lint-free cloth that has been
slightly moistened with solvent.

(2) Inspect all parts for cracks, distortion, damaged threads or other visible
damage.

(3) Repair of cracked nylon fuel tank cannot be accomplished. Repair of
fuel tank assembly is by the replacement of defective components.

d.<u>Assembly</u>

(1) If removed, attach valve cap and retainer assembly (16) to frame with screw

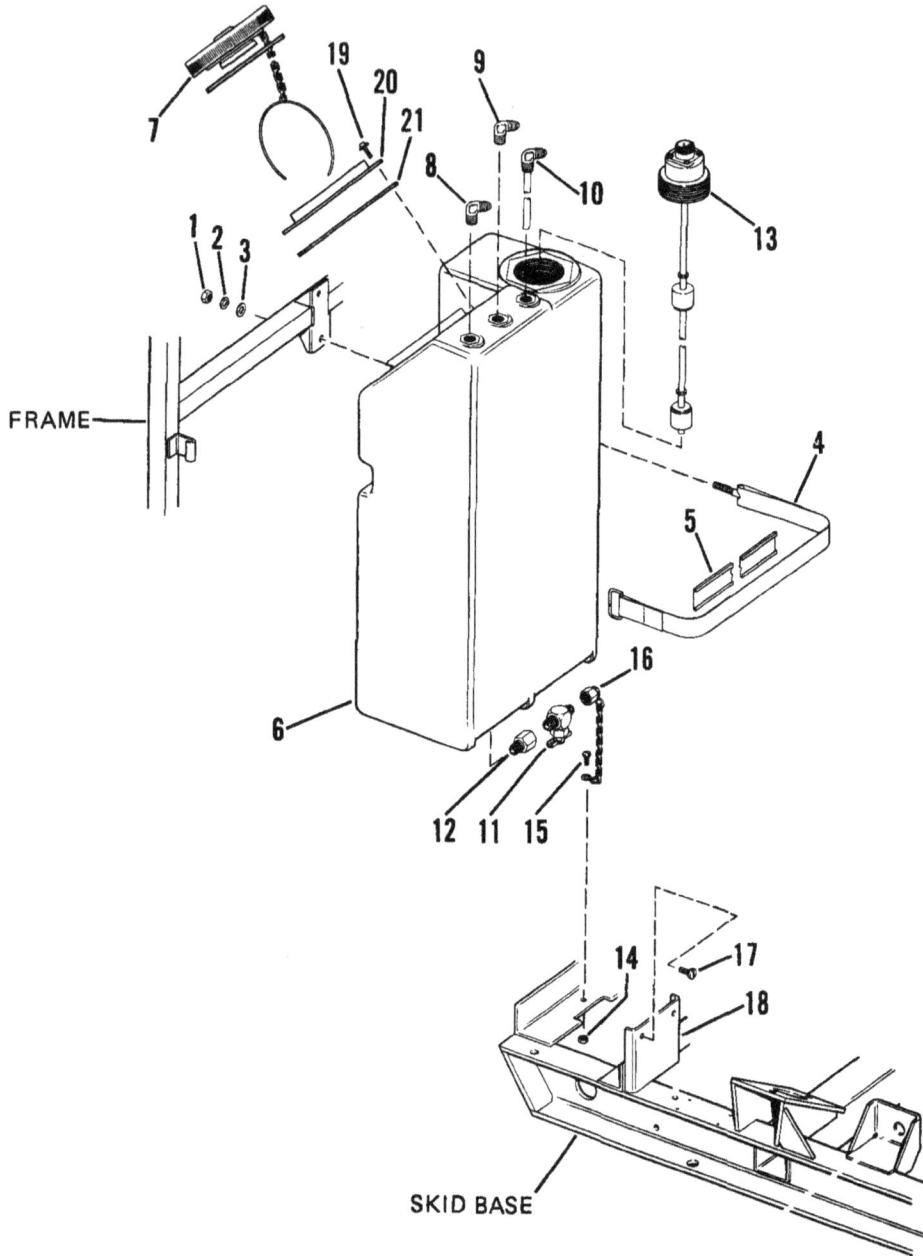

1. NUT	8. FITTING	15. SCREW
2. LOCKWASHER	9. FITTING	16. CAP AND RETAINER ASSY
3. WASHER	10. FITTING	17. SCREW
4. STRAP	11. VALVE	18. BRACKET
5. GASKET	12. FITTING	19. SCREW
6. TANK	13. SWITCH	20. FLANGE
7. CAP	14. NUT	21. GASKET

Figure 6-1. Fuel Tank.

(2) Apply teflon tape to threads of level switch (13).
(13) into fuel tank.   Install level switch

(3) Apply teflon tape to drain valve (11) and fitting (12) threads.   Install drain
valve (11) with fitting (12).

(4) Apply teflon tape to fittings (8, 9 and 10).   Install fittings (8, 9 and 10) on fuel tank.

Place gasket (21) in position and install flange (20) with screws (19).

(5)
Install filler cap (7).

(6)

e.   Testing

(1) Plug all but one of the fuel line openings on fuel tank.

(2) Submerge fuel tank in water and pressurize to 3 psig with air through the remaining fuel line fitting.

Hold pressurized tank under water for 3 minutes, no leakage of air should be visible.

(3)

Repair any visible leaks that appear at fittings by removing fitting, applying new teflon tape, and reinstalling fitting.

(4)

f. Installation

(1) Place tank (6) in position with strap (4) and gasket (5).

(2) Secure strap (4) with washer (3), lockwashers (2), and nut (1).

(3) Secure bracket (18) to frame with screw (17).

(4) Install frame, refer to paragraph 3-2.

6-3.FUEL INJECTION PUMP.

a. Removal.

WARNING

Wear protective clothing and face shield when opening fuel injection line. Fuel under high pressure may be trapped in the fuel injection line.Opening fuel injection line can cause a high pressure stream of fuel to be released which can cause severe personal injury.

CAUTION

Plug all fuel lines and connections when a fuel line is removed to prevent the entry

(1)Remove screw (1, Figure 6-2), washer (2) and clamp (3) securing injection
line (4) to push rod tube adapter (5).

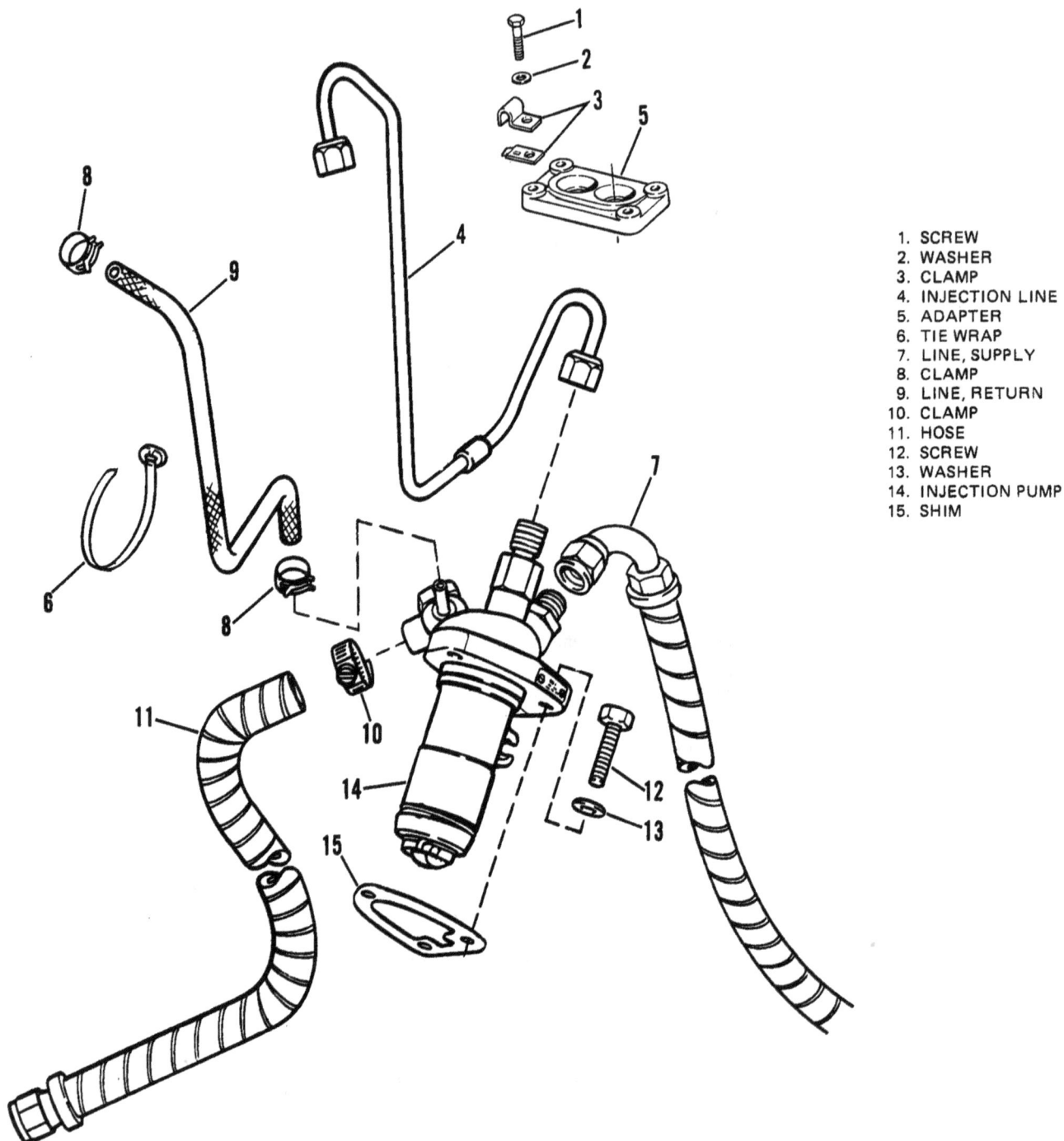

1. SCREW
2. WASHER
3. CLAMP
4. INJECTION LINE
5. ADAPTER
6. TIE WRAP
7. LINE, SUPPLY
8. CLAMP
9. LINE, RETURN
10. CLAMP
11. HOSE
12. SCREW
13. WASHER
14. INJECTION PUMP
15. SHIM

Figure 6-2. Fuel Injection Pump.

MARINE CORPS TM 05926B/06509B-34
ARMY   TM 5-6115-615-34
NAVY   NAVFAC P-8-646-34
AIR FORCE TO 35C2-3-386-32

(2)   Remove tie wrap (6) and injection line (4).

(3)   Remove supply line (7).

(4)   Remove clamps (8) and return line (9).

(5)   Remove clamp (10) and hose (11).

NOTE

If pump is pushed upwards as screws (12) are removed, rotate crankshaft approximately 90° to rotate camshaft to low point of cam lobe.

(6) Remove three screws (12) and washers (13).

(7) Position external arm of governor linkage (1, Figure 6-3) to midpoint of travel and remove injection pump (14, Figure 6-2) and shims (15).

(8) Do not remove actuator or fork unless damaged, refer to TM 05926B/06509B-12/TM 5-6115-615-12/NAVFAC P-8-646-12/TO 35C2-3-386-31 manual, paragraph 4-44.

c. Cleaning and Repair.

(1) Wipe all parts with a clean lint-free cloth that has been slightly moistened with solvent.

(2) Inspect all threaded parts for damaged threads.

(3) Inspect fuel lines for obstructions or kinks.

(4) Inspect all parts for cracks, deformation, or other visible damage.

(1) Position external arm of governor linkage (1, Figure 6-3) to midpoint of travel and place injection pump (14, Figure 6-2) and shim (15) on engine.

(2) Check that injection pump rack is caught between fork on governor linkage by holding fuel injection pump and rotating governor linkage fully counterclockwise. Rotate fuel pump and feel for engagement in both directions. Repeat with governor linkage rotated fully clockwise.

(3) Install and tighten three capscrews (12) with washers (13).

(4) Set injection pump timing, refer to step e, Adjustment.

(5) Apply sealing compound (Table 2-2, item 5) to capscrews (12) and tighten securely.
(6) Recheck injection pump timing, refer to step e, Adjustment.

(7)   Apply sealing compound to seal.   (Table 2-2, item 8) to edges of injection pump (14)

(8)   Install fuel injection   line (4).

(9)   Secure injection line (4) to adapter (5) with screw (1), washer (2) and clamp (3).

(10)   Install supply line (7).

(11)   Install new return line (9) and secure with new tie wraps (6) and clamps (8).

(12)   Install hose (11) with clamp (10).

(13)   Prime the fuel lines.
       Disconnect fuel injection line (4) at the fuel injector.

       (a)   Manually turn engine until fuel appears at the open end of the fuel injection line.

       (b)

       (c)   Connect   fuel injection line (4) to fuel injector.

e. Adjustment.

NOTE

Timing   of injection pump requires that engine timing marks be used.

(1)   Connect a high pressure timing pump (refer to Table 2-1) to the injection pump.   High pressure pump outlet is connected to injection pump supply and injection pump return line is connected to return tank on pump.
Start timing pump and rotate external linkage (1, Figure 6-3) to the fuel injection pump to the fuel off position coun-

(2)   terclockwise (4-5 mm).Fuel should   (fully clockwise). Rotate linkage be flowing from the return line.

(3)   Rotate engine slowly (clockwise, viewed observing the flow   from front flywheel) while line.

(4)   When fuel stops flowing from return line (there will be several drops of leakage), stop rotating the engine.

(5)   Rotate the engine back (counterclockwise, viewed from front flywheel) until fuel is flowing from the return line.

Very slowly rotate the engine (clockwise, viewed from front flywheel) and observe fuel return line.

(6)   Stop rotating engine when fuel stops flowing from the return line.

(7)

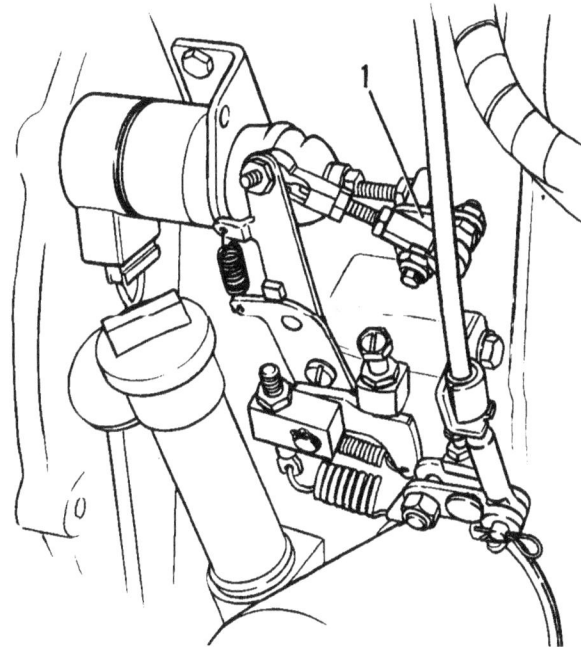

Figure 6-3.External Governor Linkage.

1. EXTERNAL GOVERNOR LINKAGE

(8) Note the number of degrees the engine is at before TDC.

(9) Injection pump timing should be at 22 degrees before TDC.

(10) Timing is retarded by adding shims (15, Figure 6-2). Timing is advanced by removing shims (15).

Shim 0.0045 in. (0.1143 mm) = 1 degree of timing

(11) Add or subtract shims (15) as required.Thickness of shims is determined by the number of small indented holes on the shim surface.

No holes: 0.0150 to 0.0165 in. (0.38 to 0.42 mm)
One hole: 0.0035 to 0.0043 in. (0.09 to 0.11 mm)
Two holes: 0.0055 to 0.0063 in.   (0.14 to 0.16 mm)
Three holes: 0.0070 to 0.0087 in.   (0.18 to 0.22mm)

6-4.   FUEL INJECTOR.

a. Removal.

WARNING

Wear protective clothing and face shield when opening fuel injection line.Fuel under high pressure may be trapped in the fuel injection line.Opening fuel injection line can cause a high pressure stream of fuel to be released which can cause severe personal injury.

(1) Disconnect fuel supply and return lines at the injector (see Figure 6-4). Plug openings in injector and lines to prevent the entry of dirt.

(2)   Remove nut (1, Figure 6-4) securing injector clamp (2). Remove clamp (2).

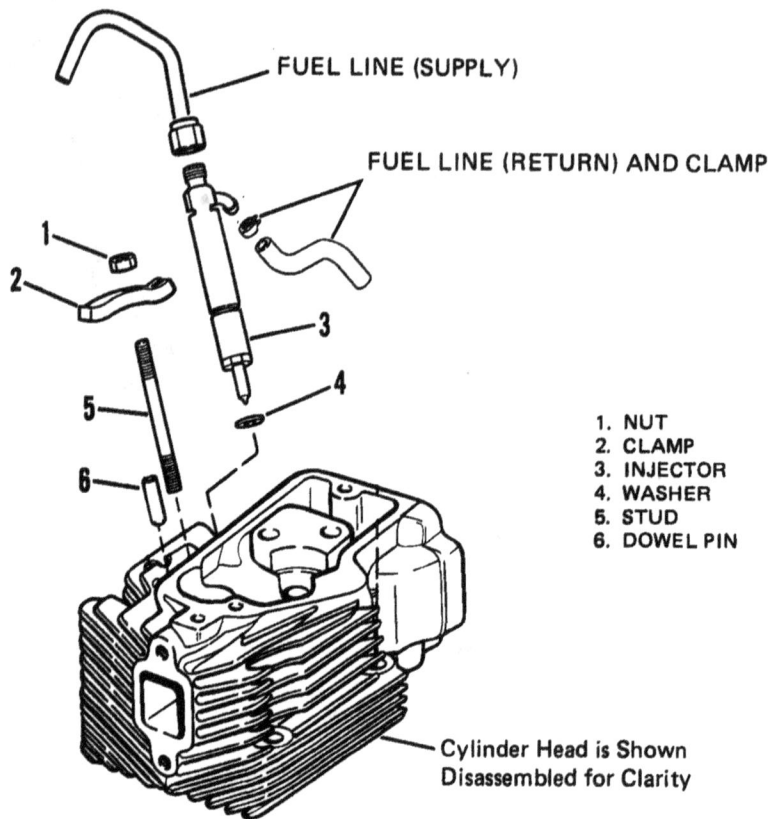

Figure 6-4. Fuel Injector Removal and Installation.

NOTE

Make sure copper washer (4) is removed from cylinder head.

(3)   Remove injector (3)   and copper washer (4). Discard washer (4).

(4)   Stud (5) and dowel   pin (6) are not removed unless visibly damaged.

(1)   Wipe parts with a clean lint-free cloth that has been slightly moistened with solvent.

(2)   Inspect parts for damaged threads, cracks, distortion or other visible damage.

(3)   Clean four spray holes of injector nozzle with cleaning needle from needle cleaning kit (Bosch P/N KDEP 1043), refer to Table 2-1.

MARINE CORPS TM 05926B/06509B-34
ARMY   TM 5-6115-615-34
NAVY   NAVFAC P-8-646-34
AIR FORCE   TO 35C2-3-386-32

c.Testing.

WARNING

Do not allow nozzle to spray against skin.Fuel under nozzle pressure can penetrate flesh
and cause a serious infection.

(1) Connect fuel injector to nozzle tester (1, Figure 6-5), (Bosch P/N 0 681
143 014), refer to Table 2-1.

Figure 6-5.Fuel Injector Testing.

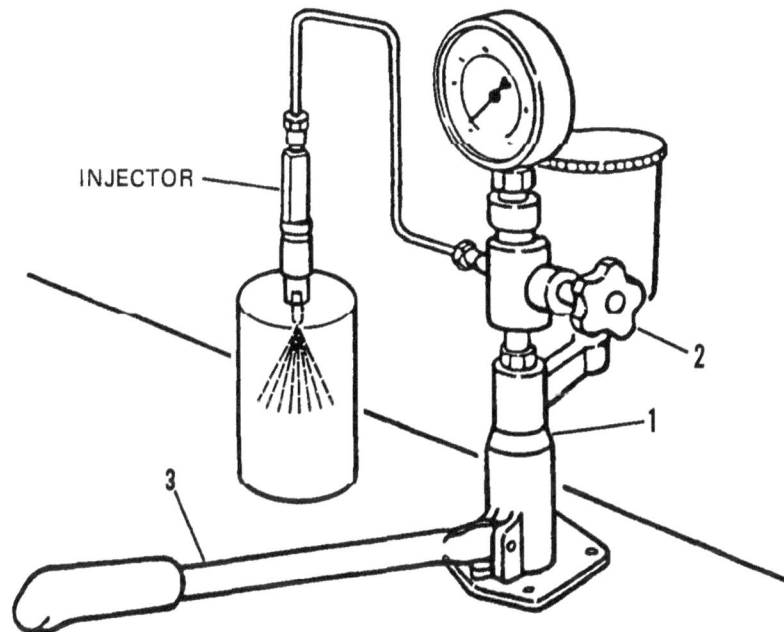

INJECTOR

1. NOZZLE TESTER
2. VALVE, GAUGE
3. LEVER, PUMP

(2) Close the tester pressure gauge valve (2). Depress the tester pump lever
(3) rapidly (6-8 strokes to bleed the system).

(3) Test opening pressure as follows:

(a) Open the tester pressure gauge valve (2). Slowly depress the pump
lever (3) of the nozzle tester.

(b)   Read the opening pressure (the pressure when the nozzle opens and the spray starts).

may be open
NOTE

If the pressure gauge valve (2)

reads erratically, the gauge too much.

(c)   Opening pressure should be between 3120 and 3260 psi (215 to 225 kPa).

(d)   Opening pressure can be adjusted by adding or removing shims, (8, Figure 6-6).Adding shims increases opening pressure, removing shims decreases the opening pressure.Refer steps d. and e. for Disassembly and Assembly procedures.

(4) Test for leakage as follows:

(a)   With the pressure gauge still open, depress the pump lever until the pressure is 50 psi (344 kPa) below the opening pressure.Maintain pressure with pump lever.

If no drops fall from the nozzle tip within 10 seconds, the nozzle is good.
(b)

If a drop falls from the nozzle within 10 seconds, the nozzle leaks and should be cleaned and retested.   If nozzle continues to drip, it should be replaced.
(c)

d.____Disassembly.

CAUTION

Disassembly and assembly of fuel injector requires a high standard of cleanliness. Small amounts of dirt can damage the injector.

CAUTION

Hold the needle valve (4) by the stem only. Skin oils will corrode the finely lapped surfaces of the needle valve.

CAUTION

The needle valve (4) and nozzle (3) are amatched set. Nozzle parts from one fuel injector cannot be interchanged with nozzle parts from another fuel injector.

CAUTION

If nozzle (3) turns when nozzle nut (2) is turned, soak the fuel injector in diesel fuel. Nozzle (3) has locating pins which may be sheared off if nozzle (3) turns with nozzle nut (2).

(1) Clamp the nozzle holder (1, Figure 6-6) in a vise. nut (2) approximately 1/4 of a turn.

Use a wrench to remove nozzle

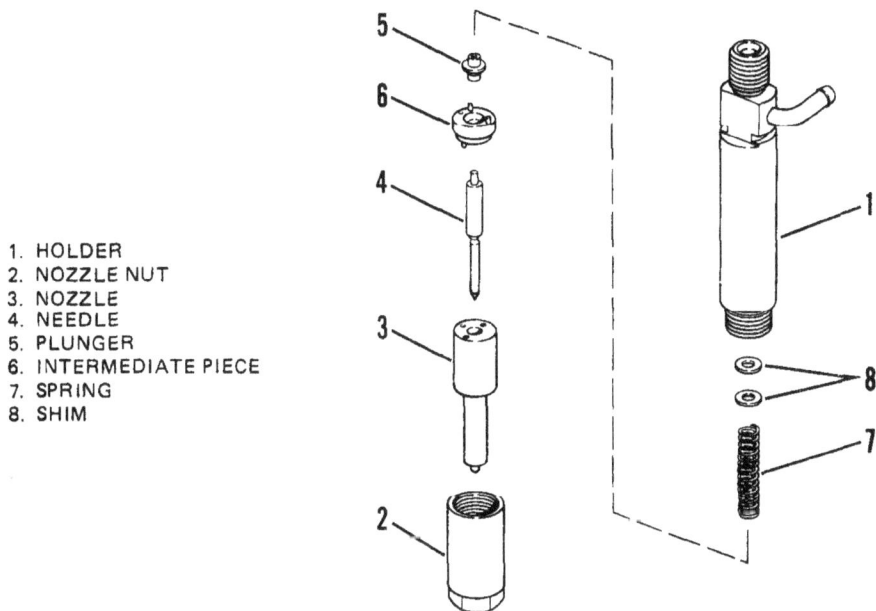

1. HOLDER
2. NOZZLE NUT
3. NOZZLE
4. NEEDLE
5. PLUNGER
6. INTERMEDIATE PIECE
7. SPRING
8. SHIM

Figure 6-6.Fuel Injector.

CAUTION

Apply downward pressure on nozzle to prevent spring (7) pressure from damaging the nozzle nut (2) threads which could contaminate the nozzle holder (1) with chips.

(2) Apply downward pressure against nozzle (3) and spring (7). (2).

Remove nozzle nut

(3) Remove nozzle (3) and nozzle needle (4) as an assembly.

(4) Remove intermediate piece (6).

(5) Remove plunger (5), spring (7) and shims (8).

e. Cleaning, Inspection, and Repair.

(1) Check nozzle needle (4, Figure 6-6) and nozzle (3) by lifting nozzle needle (4) halfway out of nozzle (3) and allowing it to drop. Needle must slide slowly and smoothly into nozzle under its own weight without sticking. If needle sticks, clean needle and nozzle in diesel fuel and repeat test. Replace both nozzle and needle if needle continues to stick.

(2) Remove any carbon residue from nozzle (3) tip and nozzle nut (2). Never use emery paper or any metal scraper to clean nozzle. Only use the hardwood

CAUTION

scrapers provided as part of the nozzle cleaning kit.

(4) Clean and polish the nozzle needle seating surface of the nozzle (3) with the hardwood scraper. Scraper is part of needle cleaning kit (Bosch P/N KDEP 1043), refer to Table 2-1.

(5) Thoroughly wet. clean all components in diesel fuel before assembly. Assemble

f. Assembly.

(1) Place nozzle (3, Figure 6-6) with nozzle needle (4) into nut (2).

(2) Place shims (8) into nozzle holder (1).

(3) Place intermediate piece (6) onto nozzle (3).Make sure that pins in intermediate piece (6) fit into holes in nozzle (3).

(4) Install plunger (5) and spring (7).

(5) Screw nozzle nut (2) onto nozzle holder (l). Tighten nut (2) to 30 ft lbs (40 N.m).

g.

Installation.
(1) If removed, apply sealing compound (Table 2-2, item 5) to stud (5, Figure 6-4) and install. If removed, install dowel pin (6).

(2) Seat new copper washer (4) in injector bore.

(4) Secure injector with clamp (2) and nut (1).Tighten nut (1) to a torque of 7 to 8 ft lbs (9 to 11 N.m).

(5) Connect fuel supply and return lines to injector.

CHAPTER 7

MAINTENANCE OF THE ENGINE

7-1. GENERAL. This chapter contains all the procedures necessary to remove, disassemble, clean and inspect, assemble, and install the engine or individual components.

7-2.ENGINE ASSEMBLY.

a.Test.

(1) Perform engine compression test.

NOTE

The compression pressure given for the compression test is based upon a normal engine starting speed.A weak battery or faulty starter may result in a lower compression reading. Make sure that the starting system is functioning normally before checking compression pressure.

Remove fuel injector, (refer to paragraph 6-4), and remove electrical connector to shutdown solenoid.

Assemble compression gage (refer to Table 2-1) and fitting (refer to Table 2-1) with sealing washer into fuel injector port.

(a)   Do not run starter any longer than 30 seconds.Allow starter to cool a minimum of two minutes after each engagement.

(b)

CAUTION

(c)   Crank engine over   until compression gage indicator stops moving. Record reading.

(d)   Remove compression gage, squirt a small amount of engine oil into the injector port, and use starter rotor to crank engine for a few seconds. Reinstall compression gage.

Crank engine over until compression gage indicator stops moving.
(e)   Record reading.

Remove compression gage and install fuel injector, refer to paragraph 6-4 for fuel injector installation.
(f)

(g) Compression pressure should be between 325 and 375 psi (2,413 and 3,100 kPa) for both the dry and wet (oil squirted on piston) test. Refer to Table 7-1 for an analysis of the compression tests.

Table 7-1. Compression Test Faults.

Indication	Probable Fault/Corrective Action
Compression pressure for both are OK./no corrective action is needed.	Valves, valve seats, piston, and piston rings wet and dry test normal
Low compression pressure for erly.adjust valves. if compression	Burned valves or seats, valves not adjusted both wet and dry test properly. pressure is still low, repair cylinder head.
Low compression pressure for worn beyond acceptable limits/replace worn (2) Perform fuel transfer pump output test.	Piston, piston rings, or cylinder sleeve is dry test and normal or near normal pressure for wet test. components.

(a) Test fuel transfer pump with procedures given in TM 05926B/06509B-12/TM 5-6115-615-12/NAVFAC P-8-646-12/TO 35C2-3-386-31 manual, paragraph 4-34.

(b) Failure of the fuel transfer pump to achieve an output pressure of 4 to 7 psi (27 to 48 kPa) indicates that the fuel transfer pump should be replaced.

b.   Removal.   Refer to paragraph 2-9.

c.Disassembly.
Remove engine, refer to paragraph 2-9.

Remove fuel cutoff solenoid, refer to TM 05926B/06509B-12/TM 5-6115-615-12/NAVFAC P-8-646-12/TO 35C2-3-386-31

(1)   manual, paragraph 4-37.

(2)   Remove sound shield and air scroll as an assembly, refer to paragraph 7-10.

(3)   Remove air flow exit duct and intake manifold, refer to TM 05926B/06509B-12/TM 5-6115-615-12/NAVFAC P-8-646-12/TO 35C2-3-386-31 manual, paragraph 4-49.

(4)   Remove starter assembly, refer to TM 05926B/06509B-12/TM 5-6115-615-12/NAVFA P-8-646-12/TO 35C2-3-386-31 manual, paragraph 4-48.

(5)   Remove governor, refer to TM 05926B/06509B-12/TM 5-6115-615-12/NAVFAC P-8-646-12/TO 35C2-3-386-31 manual, paragraph 4-44.

(6)   Remove fuel transfer pump, refer to TM 05926B/06509B-12/TM 5-6115-615-12/NAVFAC P-8-646-12/TO 35C2-3-386-31 manual, paragraph 4-34.

(7)

(8)   Remove   front and rear cylinder head wrappers, refer to paragraph 7-6.

(9)   Remove   fuel injection pump, refer to paragraph 6-3.

(10)   Remove
646-12/TO 35C2-3-386-31 manual, paragraph 4-50.

(11)   Remove   fuel injector, refer to paragraph 6-4.

(12)   Remove   cylinder head, refer to paragraph 7-6.

(13)   Remove   rocker arms and push rods, refer to paragraph 7-8.

(14)   Remove   lifters and push rod tubes, refer to paragraph 7-9.

(15)   Remove   flywheel and engine fan, refer to paragraph 7-10.

(16)   Remove   stator, refer to paragraph 7-11.

(17)   Remove   gear cover and seal, refer to paragraph 7-12.

(18)   Remove   oil pump, refer to paragraph 7-14.

(19)   Remove   camshaft, refer to paragraph 7-16.

(20)   Remove   crankshaft drive gear, refer to paragraph 7-17.

(21)   Remove   flywheel cover and gear cover back plate, refer to paragraph 7-12.

(22)   Remove   cylinder and piston, refer to paragraph 7-13.

(23)   Remove   oil pan, refer to paragraph 7-3.

(24)   Remove   connecting rod, refer to paragraph 7-15.

(25)   Remove   crankshaft, refer to paragraph 7-17.

(26)   Remove   oil filter adapter, refer to paragraph 7-18.

(27)   Remove   crankshaft main seal from flywheel housing, refer to paragraph 7-17.

d.   Assembly.

(1)   Install   crankshaft   main seal on flywheel housing, refer to paragraph 7-17.

(2)   Install   oil filter   adapter, refer to paragraph 7-18.

(3)   Install   crankshaft   (without crankshaft drive gear), refer to paragraph 7-17.

(4) Install   connecting   rod, refer to paragraph 7-15.

(5) Install   oil pan, refer to paragraph 7-3.

(6) Install   cylinder and piston, refer to paragraph 7-13.

  (7)   Install   gear cover back plate and flywheel cover, refer to paragraph 7-12.

  (8)   Install   crankshaft drive gear, refer to paragraph 7-17.

  (9)   Install   camshaft, refer to paragraph 7-16.

 (10) Install   oil pump, refer to paragraph 7-14.

 (11) Install   gear cover and seal, refer to paragraph 7-12.

 (12) Install   stator, refer to paragraph 7-11.

 (13) Install   flywheel and engine fan, refer to paragraph 7-10.

 (14) Install   lifters and push rod tubes, refer to paragraph 7-9.

 (15) Install   rocker arms and push rods, refer to paragraph 7-8.

 (16) Install   cylinder head, refer to paragraph 7-6.

 (17) Install   fuel injector, refer to paragraph 6-4.

 (18) Install   glow plug, refer to TM 05926B/06509B-12/TM 5-6115-615-12/NAVFAC P-8-646-12/TO 35C2-3-386-31 manual, paragraph 4-50.

 (19) Install fuel injection pump, refer to paragraph 6-3.

 (20) Install front and rear cylinder head wrappers, refer to paragraph 7-6.

 (21) Install fuel transfer pump, refer TM 05926B/06509B-12/TM 5-6115-615-12/NAVFAC P-8-646-12/TO 35C2-3-386-31 manual, paragraph 4-34.

 (22) Install governor, refer to TM 05926B/06509B-12/TM 5-6115-615-12/NAVFAC P-8-646-12/TO 35C2-3-386-31 manual, paragraph 4-44.

 (23) Install starter assembly, refer to TM 05926B/06509B-12/TM 5-6115-615-12/NAVFAC P-8-646-12/TO 35C2-3-386-31 manual, paragraph 4-48.

 (24) Install air flow exit duct and intake manifold, refer to TM 05926B/06509B-12/TM 5-6115-615-12/NAVFAC P-8-646-12/TO 35C2-3-386-31 manual, paragraph 4-49.

 (25) Install sound shield and air scroll as an assembly, refer to paragraph 7-10.

 (26) Install fuel cutoff solenoid, refer to TM 05926B/06509B-12/TM 5-6115-615-12/NAVFAC P-8-646-12/TO 35C2-3-386-31 manual, paragraph 4-37.

e. Installation. Refer to paragraph 2-9.

f. Adjust.

  (1) Adjust fuel injection pump timing.

    (a) Adjust fuel injection pump timing, refer to paragraph 6-3.

(b) Fuel injection pump should be set for 22 degrees Before Top Dead Center (BTDC).

(2) Adjust   valve clearance, refer to paragraph 7-8.

(3) Adjust   fuel cutoff solenoid.

  (a) Fuel cutoff solenoid should be adjusted so solenoid stops fuel injection in the de-energized position. when

  (b) Disconnect governor linkage (1, Figure 7-1).

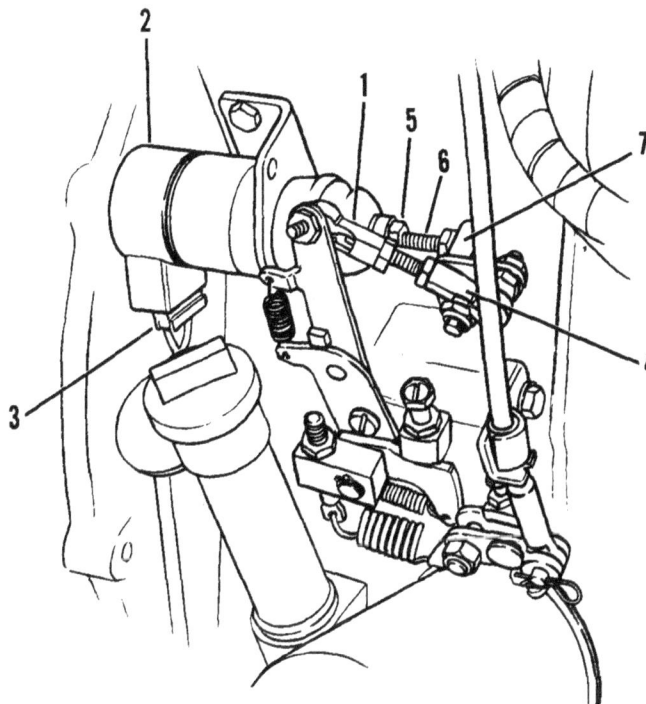

1. LINKAGE
2. SOLENOID
3. PLUG
4. ARM, EXTERNAL
5. LOCKNUT
6. SCREW, PLUNGER
7. STOP PIN

Figure 7-1.Fuel Cutoff Solenoid Adjustment.

  (c) Make sure solenoid (2) is energized.

  (d) Rotate external arm (4) fully counterclockwise.

  (e) Loosen locknut (5) and adjust plunger screw (6) for a 0.01 to 0.02 in. (0.25 to 0.51 mm) clearance between plunger screw (6) and stop pin (7) on the external linkage (4).

Tighten locknut (5) to lock plunger screw (6) position.
  (f)

(4) Adjust governor linkage.

(a)Remove nuts (1, Figure 7-2) and governor linkage (2).

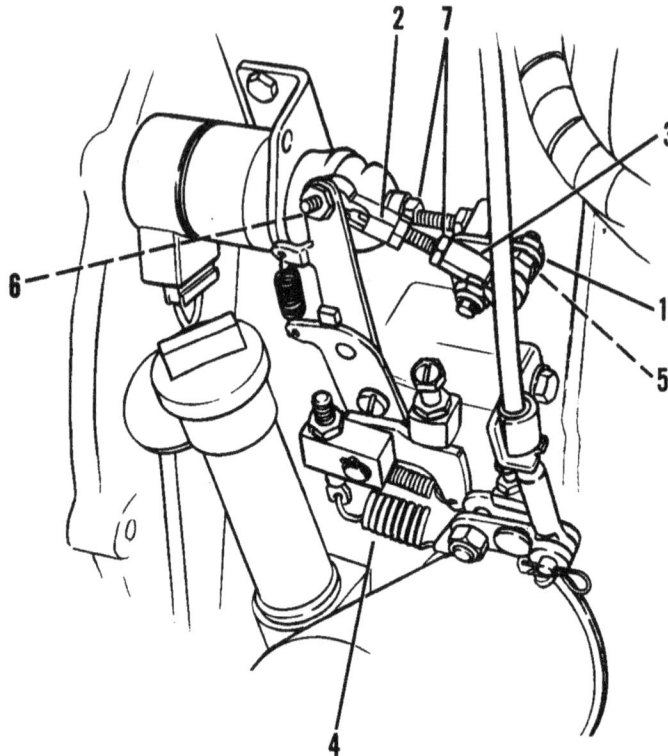

1. NUT
2. LINKAGE, GOVERNOR
3. ARM, EXTERNAL
4. GOVERNOR
5. HOLE, EXTERNAL ARM
6. HOLE, GOVERNOR ARM
7. LOCKNUTS

Figure 7-2. Governor Linkage Adjustment.

(b) Rotate external arm (3) of fuel injection pump counterclockwise towards the governor (4).

(c) Governor linkage (2) should fit easily into hole (5) in external link (3) and governor link hole (6).

(d) Adjust length of governor linkage (2) by loosening locknuts (7), turning ends in or out, and retightening locknuts (7).

(e) Secure governor linkage (2)with nuts (1).

(5) Adjust governor.

(a) Adjust governor droop, refer to TM 05926B/06509B-12/TM 5-6115-615-12/NAVFAC P-8-646-12/TO 35C2-3-386-31 manual, paragraph 4-45.

(b) Adjust governor maximum fuel screw.

1 Connect generator set to a three kilowatt (3 kw) load bank (refer to Table 2-1).

Back out high idle stop screw (1, Figure 7-3).

2

Figure 7-3.Maximum Fuel Screw Adjustment.

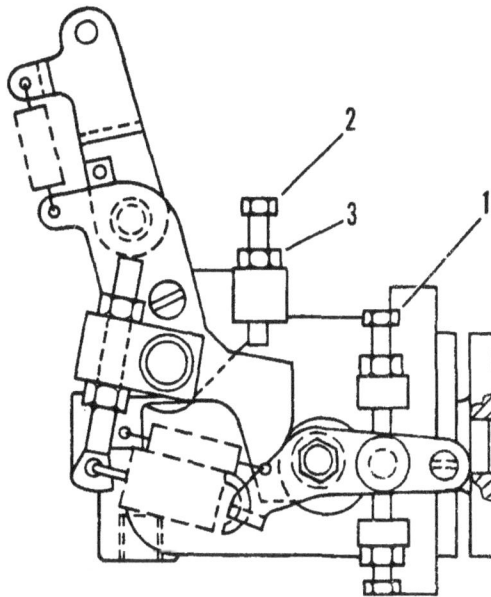

1. HIGH IDLE STOP SCREW
2. MAXIMUM FUEL SCREW
3. LOCKNUT

3 Start generator set and operate until unit reaches operating temperature.

4 Apply three kilowatt load. Adjust unit to rated speed.

NOTE

Use two wrench method when adjusting maximum fuel screw
(2).   Use a one-half inch wrench to support maximum fuel screw boss when loosening or tightening locknut
(3).

5 Loosen maximum fuel screw locknut (3) and back out maximum fuel screw
(2) counterclockwise until engine speed drops by 100 rpm (or equivalent
Hz) and tighten locknut.

<u>6</u>   Increase speed to rated rpm (or equivalent Hz) using manual speed control.

Repeat steps 4 and 5 until rated rpm is no longer obtainable.

<u>7</u>   Loosen maximum fuel screw locknut (3) and slowly turn maximum fuel screw (2) in clockwise until rated speed is obtained and tighten locknut (3).

<u>8</u>   Use manual speed control to reduce engine speed to 3400 rpm (or equivalent Hz) and switch of load.

<u>9</u>   Shut down unit.

Loosen maximum fuel screw locknut (3) and back in maximum fuel screw (2) clockwise seven-eighths of a turn and tighten locknut.

<u>10</u>   Restart unit and check and adjust no load high idle speed to 3850-3900 rpm (or

<u>11</u>

equivalent Hz).

7-3. OIL PAN.

a. <u>Removal.</u>

CAUTION

Do not apply air pressure to the engine crankcase to speed the oil drain process. Air pressure can force the oil seals out of the crankcase.

NOTE

To expedite oil draining, block unit opposite drain valve.

(1) Drain engine oil into a suitable container by opening oil drain valve (1, Figure 7-4).

(2) Remove oil drain hose (2) from oil pan.

(3) Remove screws (3) and washers (4).

(4) Remove oil pan (5).

b. <u>Cleaning, Inspection, and Repair.</u>
Scrape all traces of gasket material from both engine crankcase and oil pan.

(1)   Clean oil pan in solvent and dry.

Inspect oil pan for cracks, distortion, or other visible damage.
(2)

Repair cracked oil pan by steel welding, refer to paragraph 2-7.6.
(3)

(4)

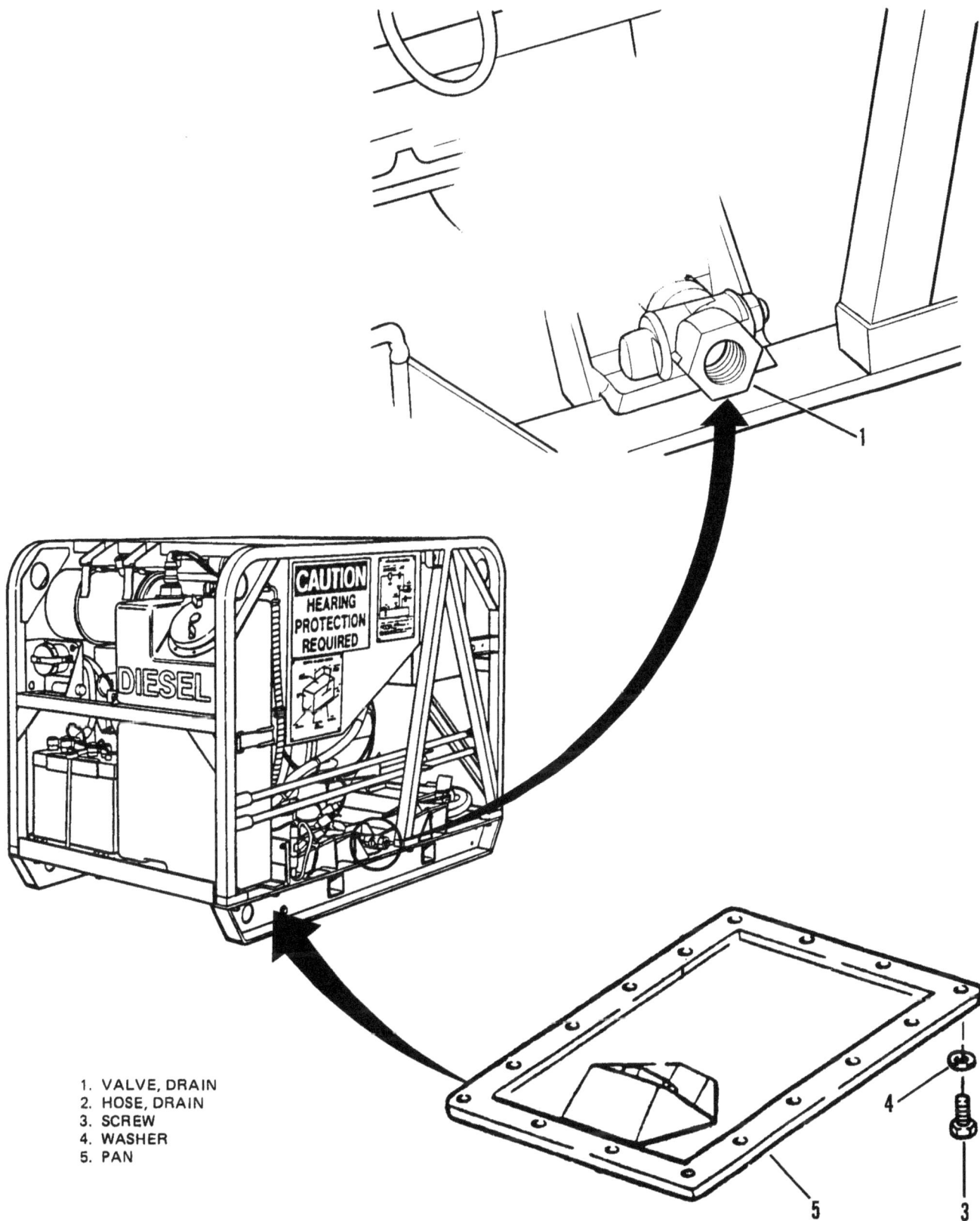

1. VALVE, DRAIN
2. HOSE, DRAIN
3. SCREW
4. WASHER
5. PAN

Figure 7-4. Oil Pan.

c.   Installation.

(1)   Apply blue sealing compound (Table 2-2, item 9) to oil pan.   Sealing compound
should be applied in a continuous 1/16 in. (1.5 mm) bead around inside of pan mounting holes.

NOTE

Drain hole on oil pan (5) should face starter side of engine.

Immediately place oil pan (5) on crankcase and align bolt holes in oil pan and crankcase.

(2)   Loosely install screws (3) and washers (4).

Tighten screws (3) in a crisscross pattern to a torque of 7 to 8 ft lbs
(3)   (10 to 11 N.m).

(4)   After assembly, allow 30 minutes minimum cure time before adding oil or running engine.

NOTE

(5)
Sealing surface must be clean and dry before applying bead.

Reinstall drain hose (2) and close oil drain valve.

Fill crankcase with the required amount and type of oil, refer to
TM 05926B/06509B-12/TM 5-6115-615-12/NAVFAC P-8-646-12/TO 35C2-3-386-31
manual, paragraph 5-3.

(6)

(7)

7-4.   STARTER ASSEMSLY.

a.   Removal.Refer to TM 35C2-3-   05926B/06509B-12/TM 5-6115-615-12/NAVFAC P-8-646-12/TO paragraph 4-8.
386-31 manual,

b.   Disassembly.

NOTE

Removal of solenoid mounting screws (1) may be difficult and require the use of an
impact type driver.

(1) Disconnect wire from solenoid.

(2) Remove two solenoid mounting screws (1, Figure 7-5) and remove fiber
washers (2) and solenoid (3).

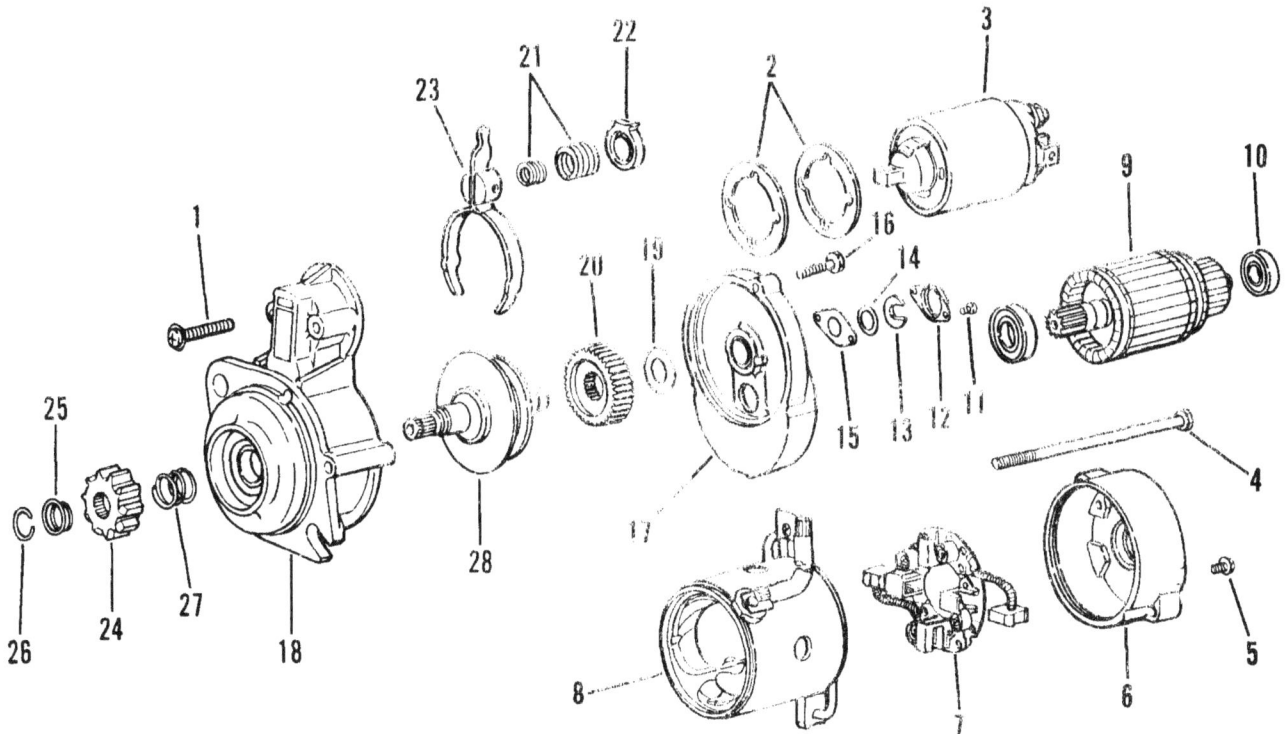

Figure 7-5. Starter Disassembly/Assembly.

1. SCREW
2. WASHER
3. SOLENOID
4. BOLT
5. SCREW
6. BRACKET (REAR)
7. BRUSH HOLDER ASSY.
8. FRAME ASSEMBLY
9. ARMATURE
10. BEARING
11. SCREW
12. COVER
13. RING, SNAP
14. WASHER
15. GASKET
16. SCREW
17. BRACKET (CENTER)
18. BRACKET (FRONT)
19. WASHER
20. GEAR
21. SPRING SET
22. PACKING
23. LEVER ASSEMBLY
24. GEAR, PINION
25. STOPPER
26. RING, RETAINING
27. SPRING
28. CLUTCH ASSEMBLY, OVERRUNNING

(3) Remove through bolts (4). Note Location of long bolt before removal. Remove brush holder retaining screws (5).

(4) Remove rear bracket (6).

(5) Remove brush holder assembly (7) and frame assembly (8).

(6) Use a screwdriver to pull brushes upward and remove frame (8). brush holder (7) from

(7) Remove armature (9) with bearing (10).

(8) Remove screws (11) and cover (12).

(9) Remove snap ring (13), washer (14) and gasket (15) from pinion shaft.

NOTE

There is a spring force behind screw (16).

(10)   Remove screw (16) that secures center bracket (17) to front bracket (18).

(11)   Count and   remove washers (19) that are used to adjust shaft end play.

NOTE

Note   the direction in which lever assembly (23) is
installed before removal.

(12)   Remove gear (20), spring set (21), packing (22) and lever assembly (23) from front bracket (18).

Push pinion gear (24) and stopper (25) down and remove retaining ring (26).

(13)   Remove pinion gear (24), stopper (25), and spring (27).

Remove overrunning clutch assembly (28).

(14)   c. Testing.

(1) Test armature for grounds.

(a) Refer to Figure 7-6.   Use a multimeter (set to read ohms) to check for
a grounded armature.

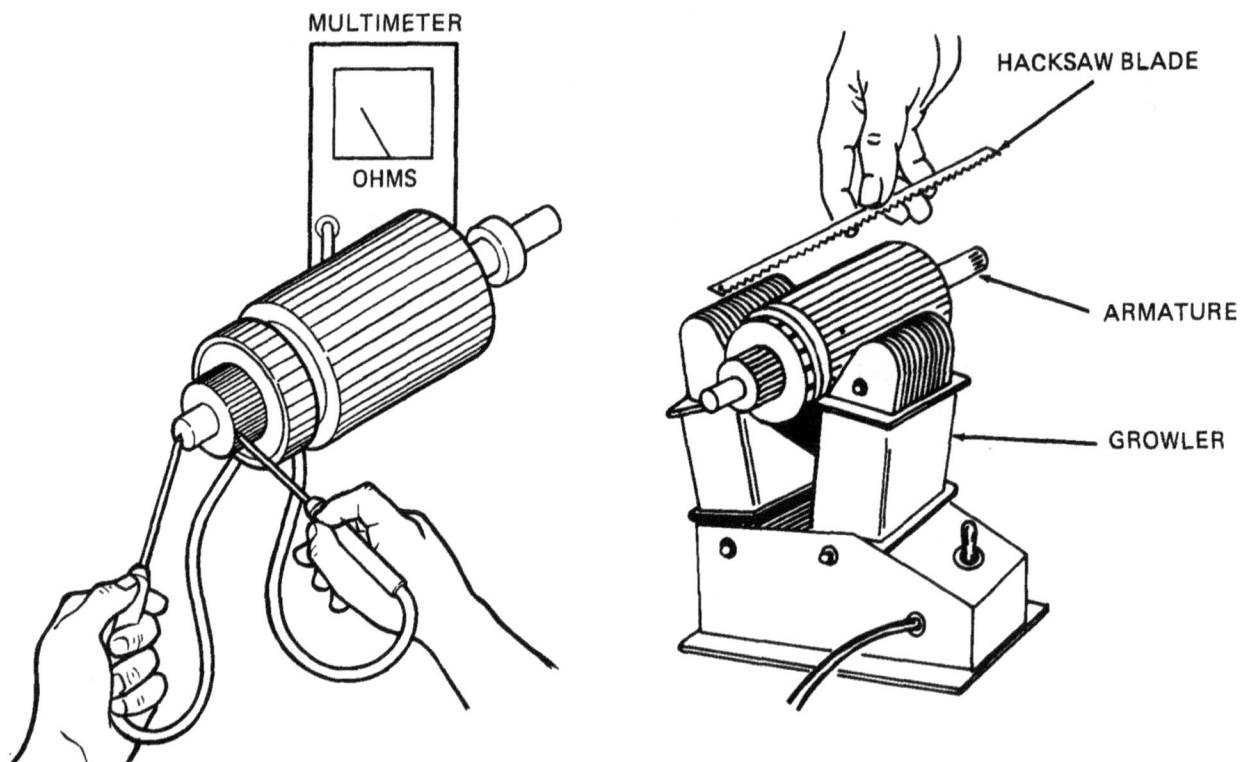

Figure 7-6.Testing Armature for Grounds or Shorts.

(b) Touch armature shaft and each commutator bar with the multimeter leads.

(c) If ohmmeter indicates a low reading (approximately 0 ohms), armature is
grounded and should be replaced.

(2) Test armature for short circuits.

(a)Use a growler (refer to Table 2-1) to locate shorts in the armature.

WARNING

Be sure to turn voltage   will be   intervals or

(b)   Refer to Figure 7-6.Place armature in growler and hold a thin steel blade (e.g. hacksaw blade) parallel and
just above the armature core while rotating armature at one-quarter turn intervals in the growler. Turn off the
growler between intervals.

(c)   A shorted armature will cause the blade to vibrate and be the core. If armature is shorted,   attracted to

(3) Test armature for an open circuit.

(a)   Refer to Figure 7-7.Use a multimeter (set to read ohms) continuity between commutator   to check for

(b)   If there is discontinuity (infinite resistance oo)the segments are
open and the armature must be replaced.

ARMATURE

MULTIMETER

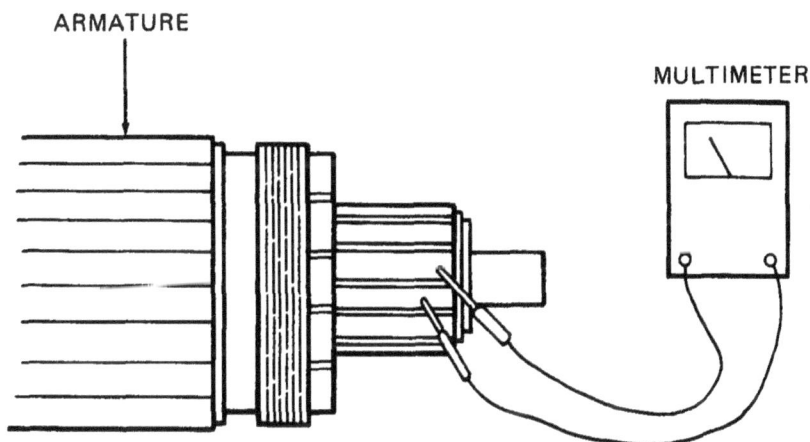

Figure 7-7. Tesing Armature for Open Circuits.

(4) Test field coil.

(a)   Usea multimeter be-   (set to read ohms) to check for continuity (0 ohms)

(b)   Ifthere is no continuity between brushes,the field coil is open and must be replaced.

      Check for continuity between field coil frame and brushes.
(c)

      If continuity exists between field coil frame and brushes, there is a ground in the field coil and field coil
(d)   should be replaced.

(5) Test solenoid.

(a)   Check for continuity between solenoid terminals "M" and "B"with a multimeter (set to read ohms).
      Refer to Figure 7-8.

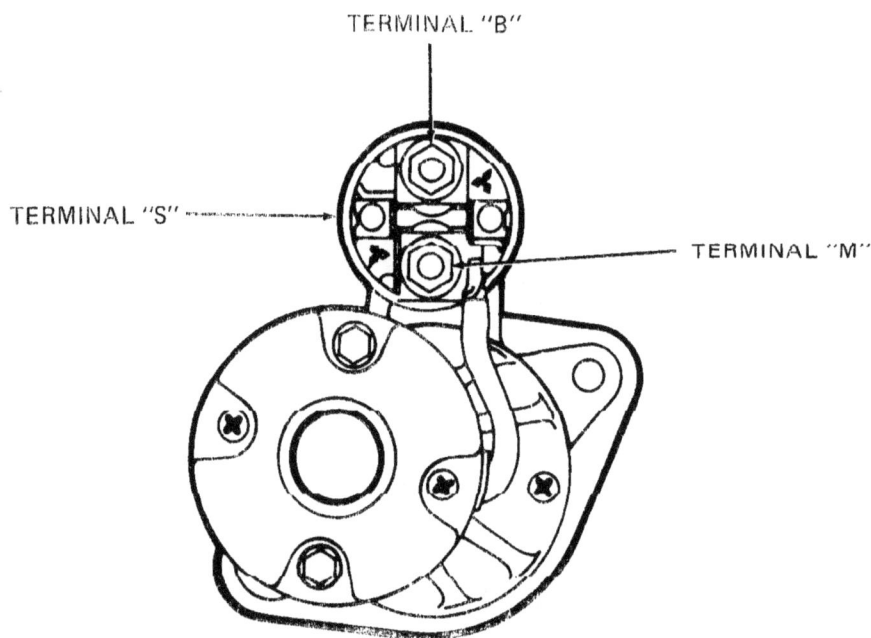

Figure 7-8. Solenoid Testing.

(b) Discontinuity should be indicated when solenoid is in its normal (out)
        position.

(c) Push solenoid plunger in and note resistance. Multimeter should indicate continuity (0 ohms). If
        there is no continuity, replace
          solenoid.

d. Cleaning, Inspection, and Repair.

<u>**CAUTION**</u>

Do not clean overrunning clutch in solvent or other cleaning solution. Washing clutch
will remove grease which may result in premature failure of the clutch.

(1) Wipe all metallic parts with a clean lint-free cloth that has been slightly dampened with solvent.

Inspect all parts for damaged threads, cracks, distortion, or other visible damage.

(2) Inspect armature commutator. If commutator is dirty or discolored, clean with number 00 or 000 sandpaper.   Use compressed air to blow sand out from between commutator segments.

(3) Clean around brushes and holders by wiping off all brush dust and dirt. Refer to Figure 7-9. If brushes are shorter than 0.453 in (11.5 mm) they should be replaced.

Figure 7-9.Brush Wear Limit.

(4) Check for the free movement of brushes.Brushes should move freely when placed in brush holders. Replace brush springs if weak or worn.

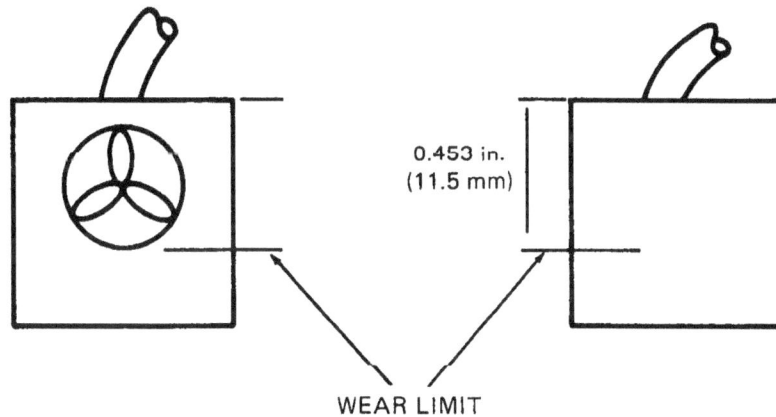

NOTE

(5) If pinion gear is worn or damaged, inspect flywheel ring gear also.

(6) Inspect armature shaft gear, bushings, reduction gear, and pinion gear for wear or damage.Replace any part that is damaged.

(7) When pinion gear is placed upon the overrunning clutch, the pinion gear should turn freely in one direction and lock when turned in the opposite direction. Refer to Figure 7-10.

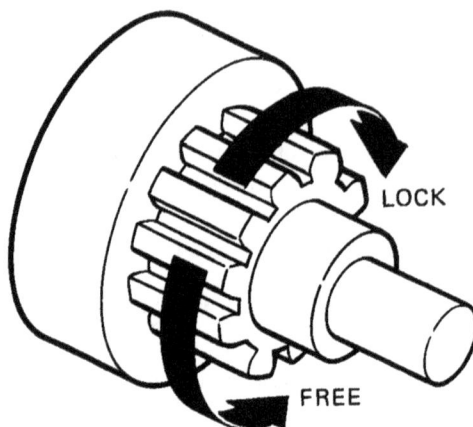

Figure 7-10. Overrunning Clutch.

e.___ Assembly.

(1) Apply a light coating of grease (Table 2-2, item 3) to the following starter components before assembling.

Armature (9) Shaft ○Gear
Reduction Gear ○(20)
Ball Bearing (10) ○
Stopper on Pinion ○Shaft (25)
Sleeve Bearing in ○Front Bracket (18)
Pinion Gear (24) ○
Lever (23) (sliding ○portion)
○

(2) Position overrunning clutch assembly (28, Figure 7-5) into front bracket (18).

(3) Install spring (27) and gear (24) on pinion shaft.

(4) Refer to Figure 7-11 and slide stopper onto pinion shaft and install retaining ring (26, Figure 7-5) in groove. Stopper (25) must fully engage retaining ring when installed.

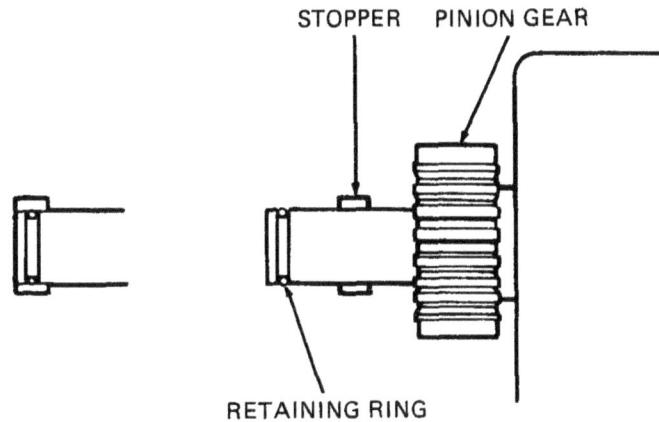

Figure 7-11. Pinion Gear Installation.

(5) Install lever assembly (23), spring set (21), and packing (22), refer to Figure 7-13.

(6) Adjust pinion shaft end play as follows:

  (a) Place reduction gear (20) on pinion shaft.

  (b) Install center bracket (17) and secure with screw (16).

  (c) Refer to Figure 7-12 and measure pinion shaft end play with a feeler gage between the center bracket

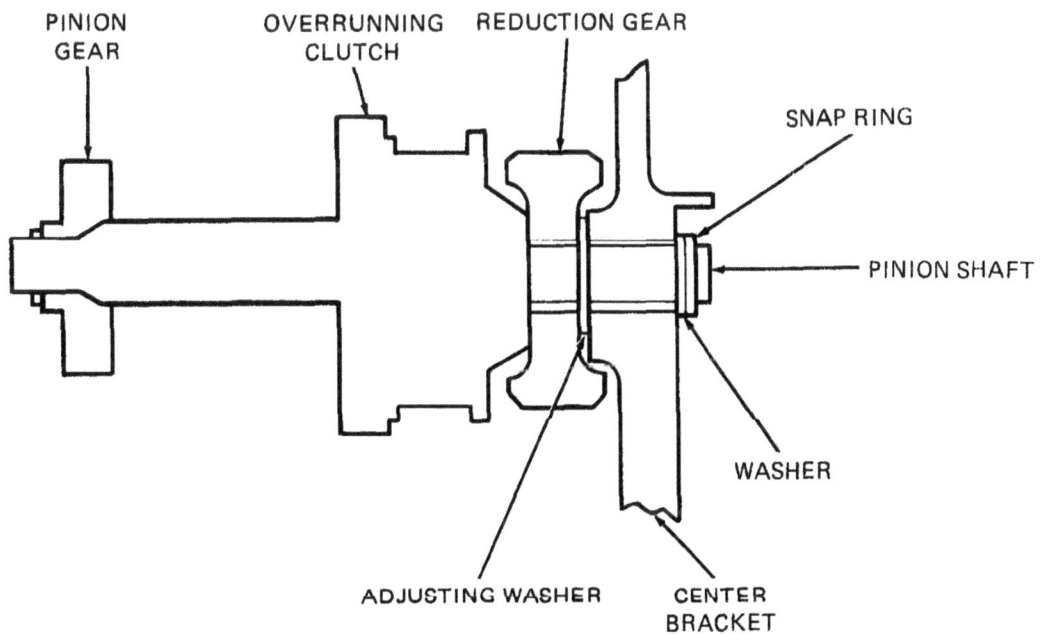

Figure 7-12.Adjusting Pinion Shaft End Play.

(d) Adjust end play to between 0.0039 to 0.0315 in (0.1 to 0.8 mm) with adjustment washers (19).

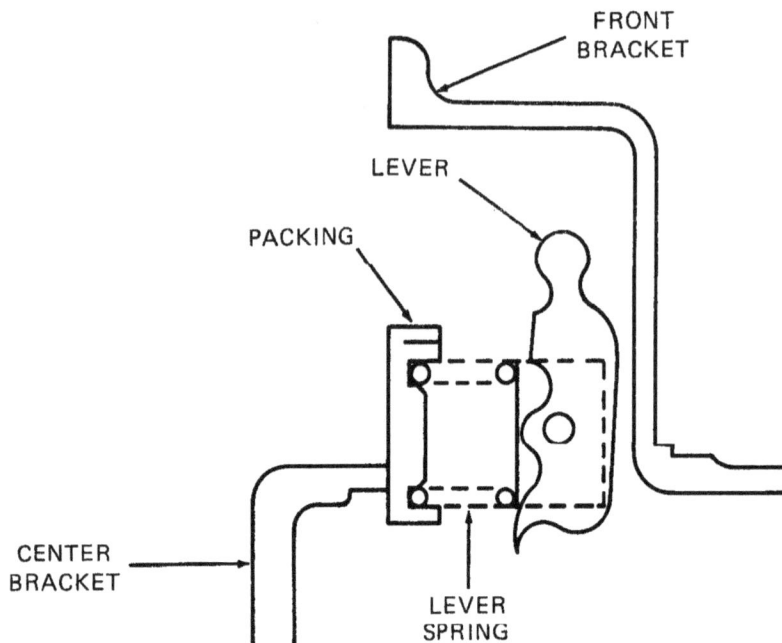

Figure 7-13. Lever Installation.

(7) Install gasket (15, Figure 7-5), washer (14) and snap ring (13).

(8) Install cover (12) with screws (11).

(9) Install armature (9) on center bracket (17).

(10) Install brush holder (7) in frame assembly (8) and position frame assembly on center bracket (17). Tab on frame assembly (8) must fit in notch in center bracket (17) at assembly.

(11) Pull brushes upward and install armature assembly (9) with bearing (10).

(12) Secure rear bracket (6) with screws (5) and bolts (4).

NOTE

Be sure that rectangular hole in solenoid plunger shaft engages lever in housing.

Apply sealing compound (Table 2-2, item 7) to mounting screws before assembly.

(13) Install solenoid (3), fiber washers (2), secure with mounting screws (1).

7-18

(14) Reconnect wire to solenoid.

(15) Check pinion gap adjustment.

    (a) Connect a 24 VDC power source as follows:

       1 Positive (+) connection to "S" terminal on solenoid (see Figure 7-8). 2 Negative (-) connection to starter body. With power applied to solenoid, starter pinion gear will shift.

    (b) Gently push the pinion shaft towards the solenoid and measure the amount of travel. Refer to Figure 7-14.

    (c)

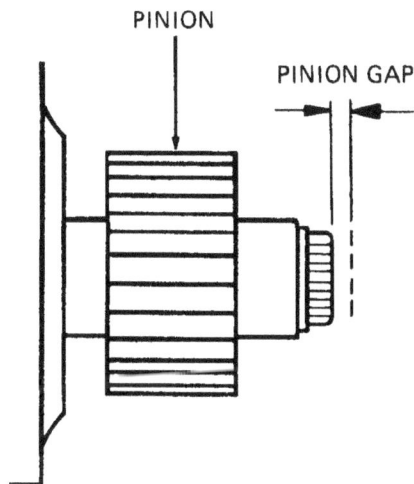

Figure 7-14. Pinion Gap Adjustment.

    (d) The pinion gap should be 0.0118 to 0.787 in. (0.3 to 2.0 mm).

    (e) Adjust the gap by increasing or decreasing the number of fiber washers (Figure 7-12). Increasing number of washers decreases the clearance and decreasing the number of washers increases the clearance.

f. Installation.   Refer to TM 05926B/06509B-12/TM 5-6115-615-12/NAVFAC P-8-646-12/TO 35C2-3-386-31 manual, paragraph 4-48 for starter installation instructions.

7-5. STARTER SOLENOID.

a. Removal.

    (1) Tag and disconnect all wires connected to the solenoid.

MARINE CORPS TM 05926B/06509B-34
ARMY   TM 5-6115-615-34
NAVY   NAVFAC P-8-646-34
AIR FORCE   TO 35C2-3-386-32

(2) Remove screws (1,Figure 7-15) that hold the solenoid (2) to the starter motor.   Remove solenoid and remove gaskets (3).

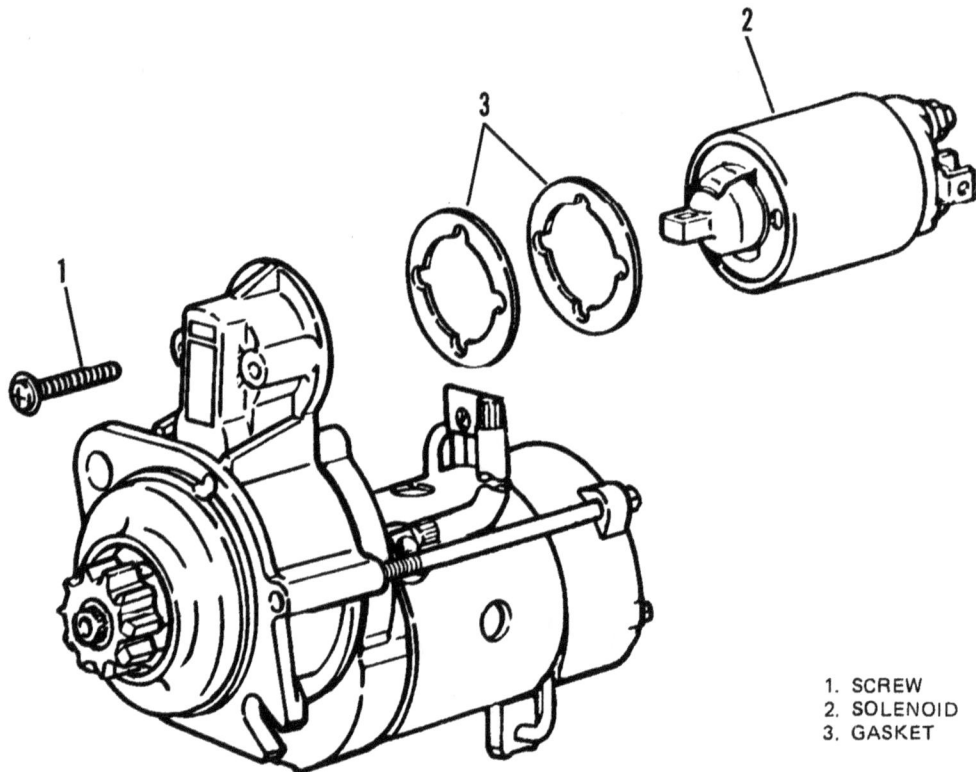

1. SCREW
2. SOLENOID
3. GASKET

Figure 7-15. Starter Solenoid.

b.Installation.

(1) Put gaskets (3) and solenoid (2) in position on the starter motor and install screws (1).

(2) Using tags as identification,connect all wires to the solenoid.

7-6.CYLINDER HEAD.

a.Removal.

(1)   Remove intake manifold, refer to TM 05926B/06509B-12/TM 5-6115-615-12/NAVFAC P-8-646-12/TO 35C2-3-386-31 manual, paragraph 4-49.

(2)   Remove muffler, refer to TM 05926B/06509B-12/TM 5-6115-615-12/NAVFAC P-8-646-12/TO 35C2-3-386-31 manual, paragraph 4-54.

(3)   Remove lifting eyes and wrappers.

(a) Remove screw (1,Figure 7-16), washer (2) and lifting eye (3).

(b) Remove nut (4), washer (5), lifting eye (6), stud (7) and upper muffler bracket (27).

7-20

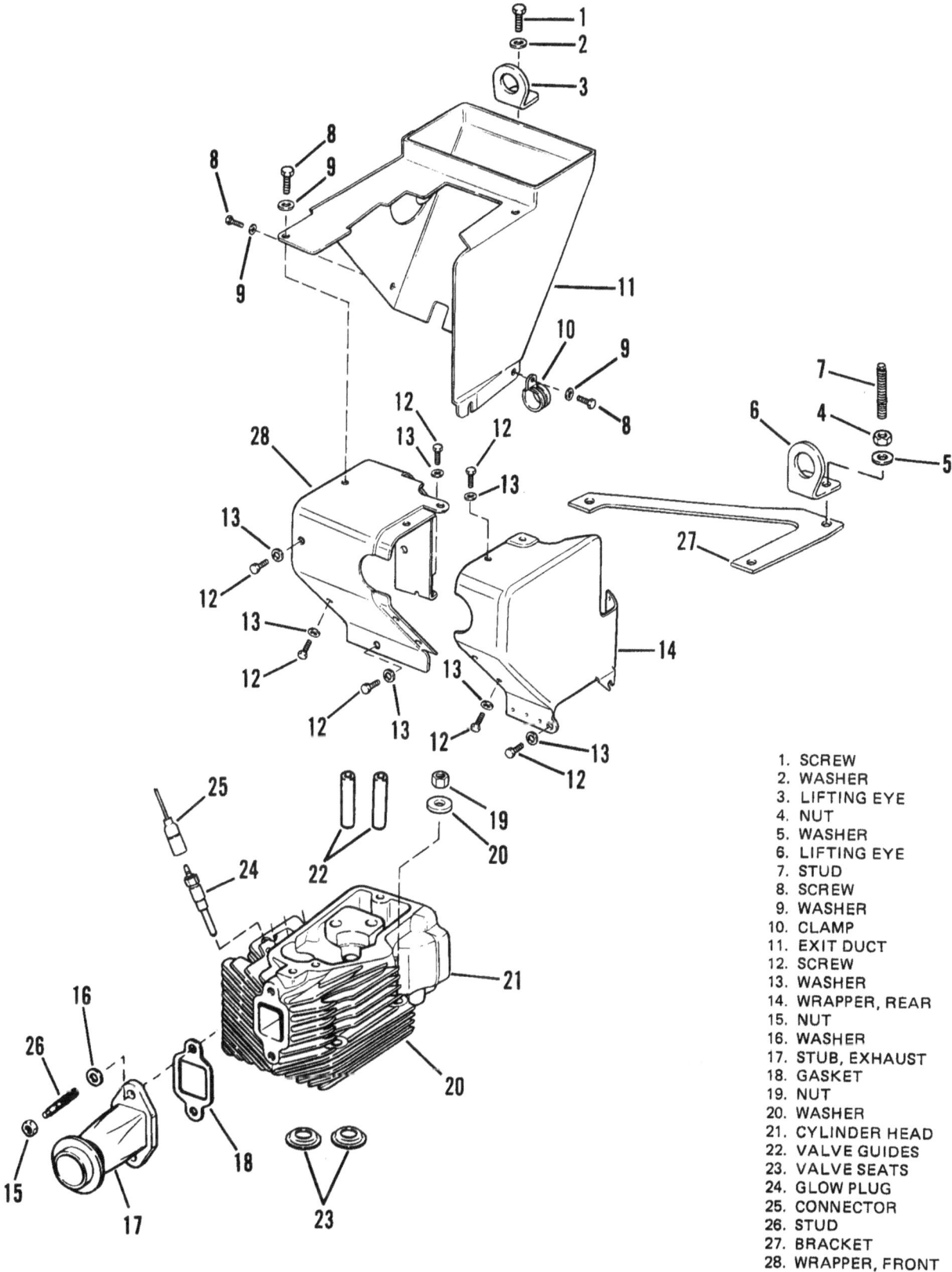

1. SCREW
2. WASHER
3. LIFTING EYE
4. NUT
5. WASHER
6. LIFTING EYE
7. STUD
8. SCREW
9. WASHER
10. CLAMP
11. EXIT DUCT
12. SCREW
13. WASHER
14. WRAPPER, REAR
15. NUT
16. WASHER
17. STUB, EXHAUST
18. GASKET
19. NUT
20. WASHER
21. CYLINDER HEAD
22. VALVE GUIDES
23. VALVE SEATS
24. GLOW PLUG
25. CONNECTOR
26. STUD
27. BRACKET
28. WRAPPER, FRONT

Figure 7-16. Clinder Head.

(c)   Remove screws (8) washers (9), clamp (10) and exit duct (11).

(d)   Remove screws (12), washers (13) and rear wrapper (14).

(e)   Remove nuts (15), washers (16), exhaust stub (17) and gasket (18), and front wrapper (28).

(f)   Remove intake manifold, refer to TM 05926B/06509B-12/TM 12/NAVFAC P-8-646- 5-6115-615-4-48. 12/TO 35C2-3-386-31 manual, paragraph

(3)   Remove rocker arms and push rods, refer to paragraph 7-8.

(4)

(5)   Remove glow plug (24), refer to TM 05926B/06509B-12/TM 5-6115-615-12/NAVFAC P-8-646-12/TO 35C2-3-386-31 manual, paragraph 4-50.

(6)   Remove four nuts (19, Figure 7-16) and washers (20).

(7)   Remove cylinder head (21).

b. Disassembly.

(1) Remove valves, refer to paragraph 7-7.

(2) Valve guides (22)and seats (23) are not removed unless indicates that they cannot be   inspection reused.Refer to step c, Cleaning and Inspection.   For removal, refer to step d,

Repair.

c. Cleaning and Inspection.

(1)   Wipe parts with a clean lint-free cloth that has been moistened solvent.   with

(2)   Remove carbon deposits with a hardwood chisel or steel brush, being careful not to scratch the valveor cylinder seating surfaces.

(3)   Measure valve guide (22) to valve clearance.For maximum allowable clearance, see Table 1-1.

If proper clearances cannot be obtained by replacing valves, replace valve
(4)   guides refer to step d, Repair.

Measure valve seat (23) width (refer to Table 1-1).   If seat width cannot be narrowed
(5)   by grinding, replace valve seat, refer to step d, Repair.

Inspect valve seats for signs of pitting, seats may be reground.   Refer to step d,
(6)   Repair.
Inspect all parts for cracks, distortion, or other visible damage.

(7)

d.   Repair.

WARNING

Do not use a punch, prybar, or chisel to remove valve seat.   Seat is made of a hardened material which may shatter and cause personal injury.

(a)   Remove valve seats (23) without damaging head.

(b)   Check the valve seat bore for burrs, cracks or rough edges. If burrs or rough edges are present, they should
be removed.

(c)   Place the new insert into a container of dry ice.   Insert must be cooled to
approximately -70°F (-57°C).
Wear protective gloves when handling dry ice and cooled valve seat.   Personal injury may

---
WARNING
---

result if the seat or ice comes in contact with unprotected skin.

Heat the cylinder head to 350° F (177° C).

(d)   Quickly remove the valve seat from the dry ice and press into the cylinder head.

(e)

(2) Grind valve seats.

CAUTION

Only remove enough material to produce a smooth seat surface.

(a)   Make sure that the valve guide is clean and apply a light coating of engine oil to the valve seat.

(b)   With a seat grinder (refer to Table 2-1), lightly grind the valve seat surface to produce a smooth seat surface.
Intake valve seat and exhaust valve seat dimensions are listed in Table 1-1.
(3) Replace valve guides.

(a) Press old valve guide (22)out from the underside of the cyinder head.

(b) After valve guide has been removed,check the bore for burrs, cracks, or rough edges.   If burrs or rough edges are present, they should be removed.

Wear protective gloves when handling dry ice and cooled valve seat.Personal injury may result if the seat or ice comes in contact with unprotected skin.

WARNING

Place the new valve guide into a container of dry ice. Guide must be cooled to approximately -70°F (-57° C).

Heat the cylinder head to 350° F (177° C).

(c) Use a mandrel to press the valve guide to between 0.681 and 0.697 in.

(d)

(e)
(17.3 to 17.7 mm) above the valve spring seat surface.

e. Assembly. Install valves, refer to paragraph 7-7.

f.   Installation.

(1) Place cylinder   head (21) onto cylinder   head studs.

(2) Make sure push   rod tubes arecorrectly   positioned in cylinder head.

(3) Apply engine oil to cylinder ensure proper   head studs, washers (20),and nuts (19) to washers (20)and nuts
    torque.Place studs and tighten lightly by   (19) on cylinder head hand.   Ensure that cylinder head is
    squarely on cylinder barrel.

(4) Tighten cylinder head nuts (19) in a crisscross pattern. Torque in three steps; 15, 26, and 33 ft lbs (20, 35, and 45 N.m). Refer to Figure 1-1.

(5) Install injector, refer to paragraph 6-4.

(6) Install rocker arms and push rods, refer to paragraph 7-8.

(7) Install glow plug (24).   Refer to TM 05926B/06509B-12/TM 5-6115-615-12/NAVFAC P-8-646-12/TO 35C2-3-386-31 manual, paragraph 4-50.

(8) Install lifting eyes and wrappers.

(a) Apply sealing compound (Table 2-2, item 7) to all screws.

NOTE

Do not tighten screws until all have been started.

(b) Place front wrapper (28)in position and secure with screws (12) and washers (13).

(c) Place gasket (18) and exhaust stub(17) in position on cylinder head and install washers (16) and nuts (15). Tighten nuts to a torque of 16-18 ft lbs (21-23 N.m).

(d) Place rear wrapper (14) in position and secure with screws (12) and washers (13).

(e) Install exit duct (11) and secure withy screws (8) and washers (9), and clamp (10).

(f) Secure lifting eyes(6) and bracket (27) with washer (5), stud (7) and nut (4).

(g) Secure lifting eye (3) with screw (1) and washer (2).

(9) Install muffler, refer to TM 05926B/06509B-12/TM 5-6115-615-12/NAVFAC P-8-646-12/TO 35C2-3-386-31 manual, paragraph 4-54.

(10) Install intake manifold, refer to TM 05926B/06509B-12/TM 5-6115-615-12/NAVFAC P-8-646-12/TO 35C2-3-386-3 manual, paragraph 4-50.

7-7. **VALVES.** a. Remov-

al.

(1) Remove cylinder head, efer to paragraph 7-6.

(2) Compress valve springs (1,Figure 7-17) with a valve spring compressor (refer to Table 2-1).

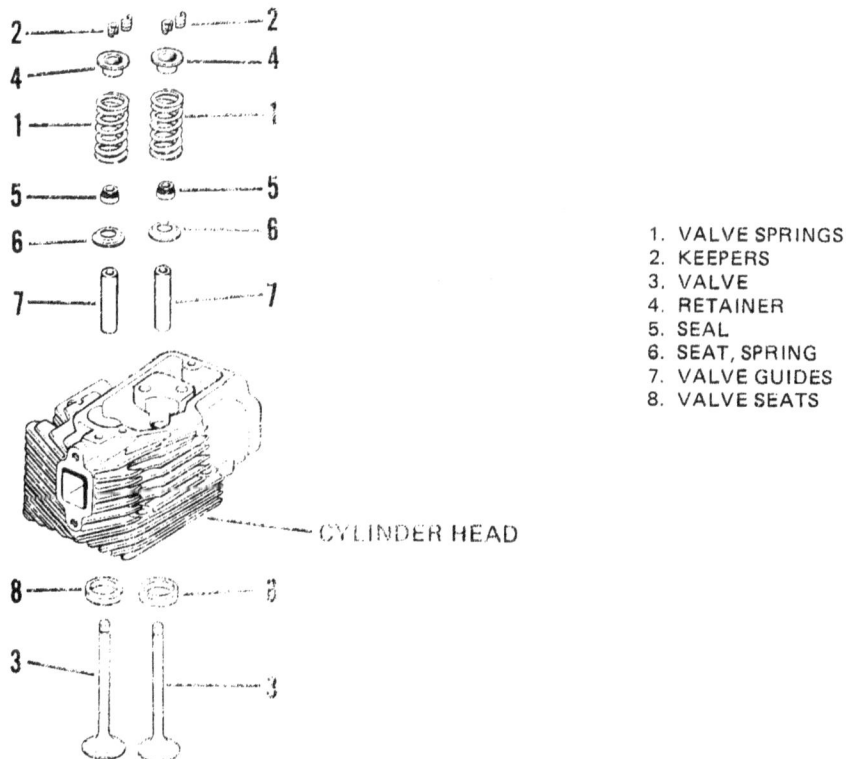

1. VALVE SPRINGS
2. KEEPERS
3. VALVE
4. RETAINER
5. SEAL
6. SEAT, SPRING
7. VALVE GUIDES
8. VALVE SEATS

Figure 7-17. Valves.

(3)   Remove valve keepers (2).

(4)   Release valve spring compressor slowly.

(5)   Remove valves (3), spring retainers (4) and valve springs (1).

(6)   Remove valve stem seals (5) and spring seats (6).

CAUTION

Intake valve is slightly larger than the exhaust valve.
Do not interchange valves.

(1)   Install valve into valve grinding tool (refer to Table 2-1).

(2)   Make sure that grinding wheel of grinding tool is well dressed before grinding valve.

(3)   Adjust the valve chuck angle to the desired degree.Make sure that valve head is close as possible to chuck to prevent vibration or bending.

(4)   Check the valve runout; if runout is greater than 0.002 in. (0.050 mm), relocate the valve in the chuck and try again.

(5)   Lightly cut across the valve face.This will reveal warpage of the valve head which may not have been apparent earlier.

(6)   To avoid overheating, make only light and slow cuts.

(7)   Grind valves to within acceptable limits, refer to Table 1-1.

c.Installation.

(1)   Install spring seats (6) over valve guides (7) into recess in cylinder head.

(2)   Press stem seals (5) into place.

(3)   Lubricate valve (3) stems and insert valves.

(4)   Install valve springs (1) and retainers (4).

(5)   Compress valve springs with a valve spring compressor.

(6)   Install keepers (2) and slowly release valve spring compressor.

(7)   Check that keepers (2)   are seated properly.

(8)   Install cylinder head,   refer to paragraph 7-6.

7-8.   **ROCKER ARMS AND PUSH RODS.**

NOTE

Exhaust rocker arm is longer than the intake rocker arm.

a. Removal.

   (1) Remove rocker arm cover:

      (a) Move clamps (1, Figure 7-18) so that breather hose (2) can be removed.

      (b) Remove breather hose (2) from connector on crankcase and breather tube on rocker arm cover (5).

      (c) Remove two screws (3),washers (4), rocker arm cover (5), and gasket
      (6).
          Discard gasket (6). rocker arm

   (2) Remove
          nuts (7).

   (3) Remove
          special washers (8), rocker arm balls (9), and rocker arms (10).

   (4) Remove
          rocker arm studs (11) and push rod plate (12).

   (5) Remove

b. Cleaningand Inspection.

   (1)   Wipe all parts with a clean lint-free cloth that has been slightly moistened with solvent.

   (2)   Inspect parts for damaged damage.   threads, cracks, distortion, or other visible

          Inspect rocker arm thrust
   (3)   spots, or other visible wear.   pads and rocker arm balls for scoring, flat

c. Repair. Deep score marks or excessively worn rocker arm or rocker balls
       cannot be repaired and should be replaced.

   (1) Place push rods (13) into   push rod tubes.

   (2) Place push rod plate (12)   on cylinder head.

   (3) Apply sealing compound (Table 2-2,   item 5) to and install rocker arm
       studs (11).

   (4) Place rocker arms (10), balls (9),   and special washers (8) on rocker arm
       studs.

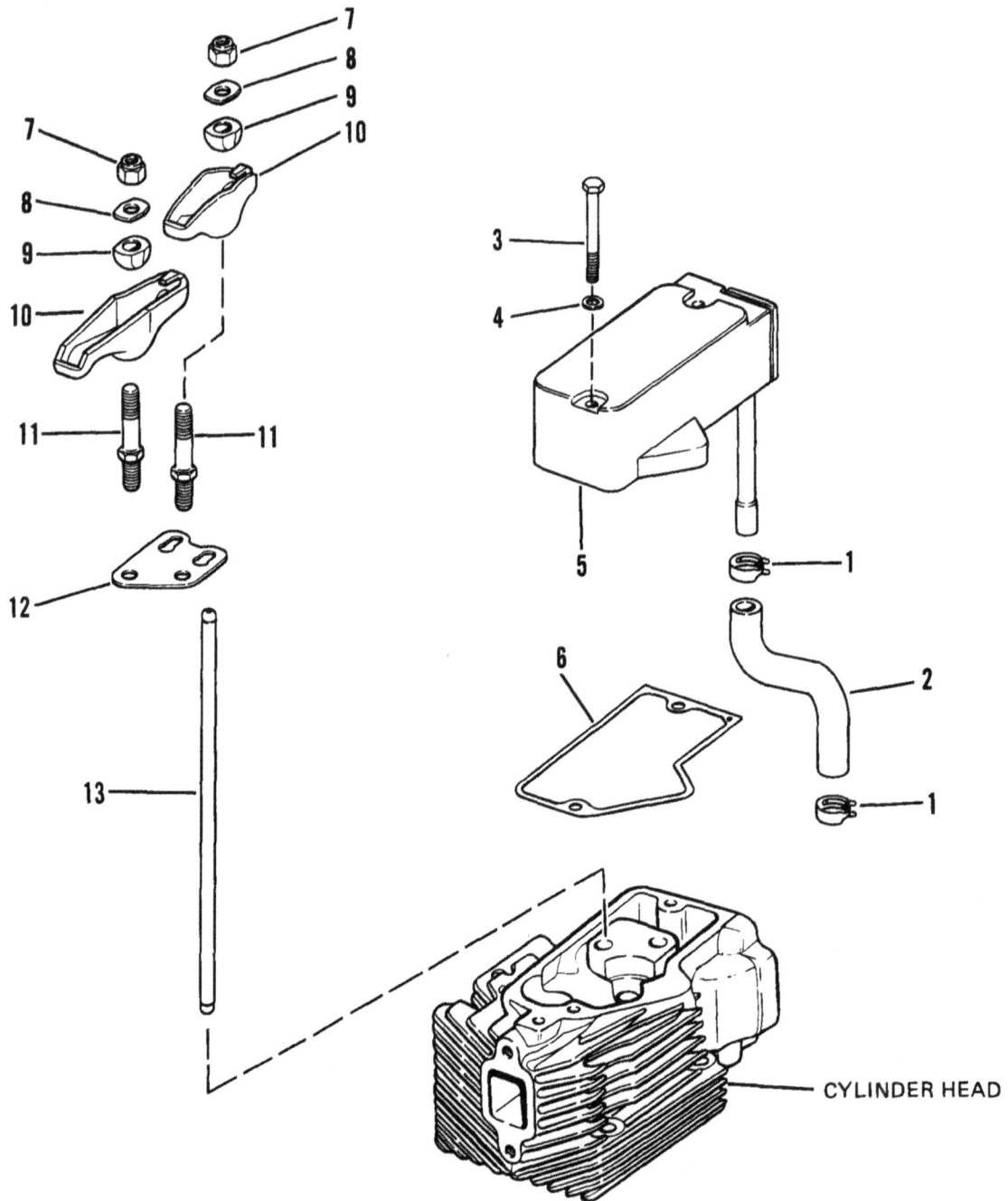

1. CLAMP	8. WASHER
2. HOSE	9. BALL
3. SCREW	10. ARM
4. WASHER	11. STUD
5. COVER	12. PLATE
6. GASKET	13. PUSHROD
7. NUT	

Figure 7-18. Rocker Arms and Push Rods.

(5) Install self-locking nuts (7). Adjust valve clearance, refer to step e, Adjustment.

(7) Lubricate rocker arms and balls liberally with engine oil.

e. Adjustment.

(1) Remove inspection plug from flywheel housing to expose timing pointer and timing marks on flywheel.The inspection plug is located on the flywheel housing between the starter motor and the engine crankcase (see Figure 7-19).

(2) Rotate engine clockwise (viewed from flywheel) until piston is at top dead center (TDC) of compression stroke (both valves closed) as indicated by the flywheel timing marks and pointer.   See Figure 7-19.

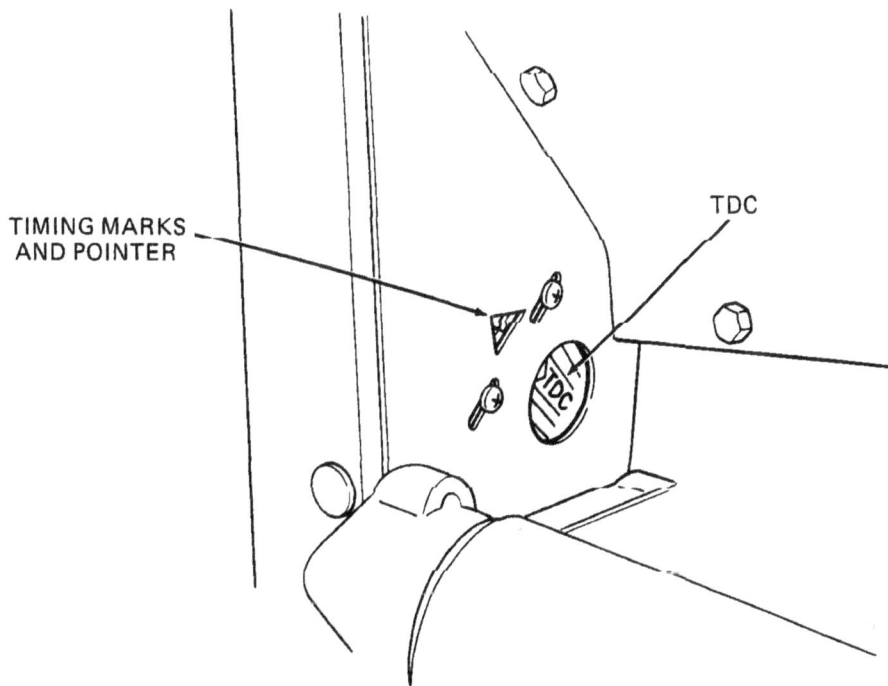

TIMING MARKS AND POINTER

TDC

Figure 7-19.Piston at Top Dead Center to Set Engine Timing.

7-29

Loosen the rocker arm retaining nuts (7, Figure 7-18) until the rocker arms move freely.Now tighten the rocker arm retaining nuts until there is 0.001 in. (0.25 mm) clearance between the valve stem and rocker arm. Use a feeler gage to measure clearance (see Figure 7-20).

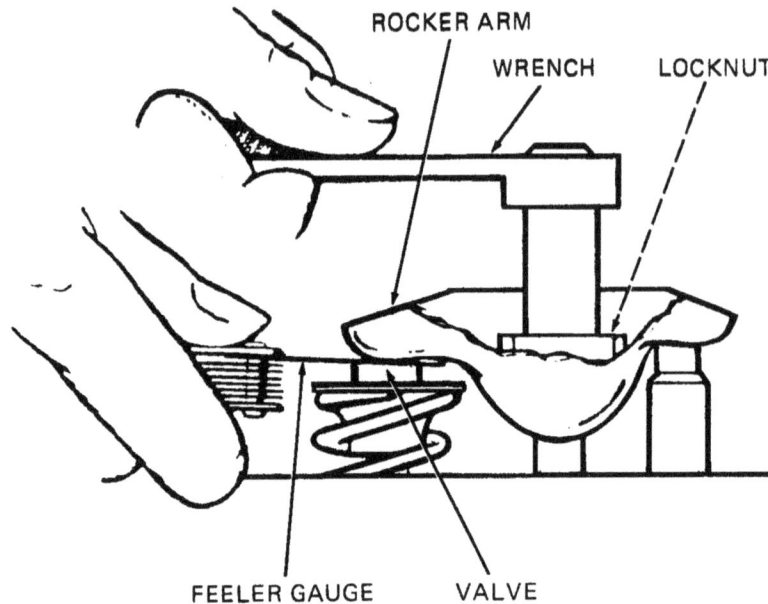

Figure 7-20. Adjusting Valve Clearance.

(4)   Tighten rocker arm retaining nuts (7, Figure 7-18) an additional 1-1/2 turns.

(5)   Allow approximately 5 minutes for lifters to leak down (drain of oil) before starting engine.

CAUTION

Never start engine   immediately after adjustment. piston causing
Valves may contact

(6)   Using a new rocker arm cover gasket (6, Figure 7-18), install rocker arm cover (5) with screws (3) and washers (4).   Be sure that breather hole in rocker cover is aligned with hole in rocker cover gasket.

7-9.   LIFTERS AND PUSH ROD TUBES.

a.   Removal.

(1) Remove cylinder head, refer to paragraph 7-6.

7-30

(2)   Remove lifter tube assembly (1, Figure 7-21).

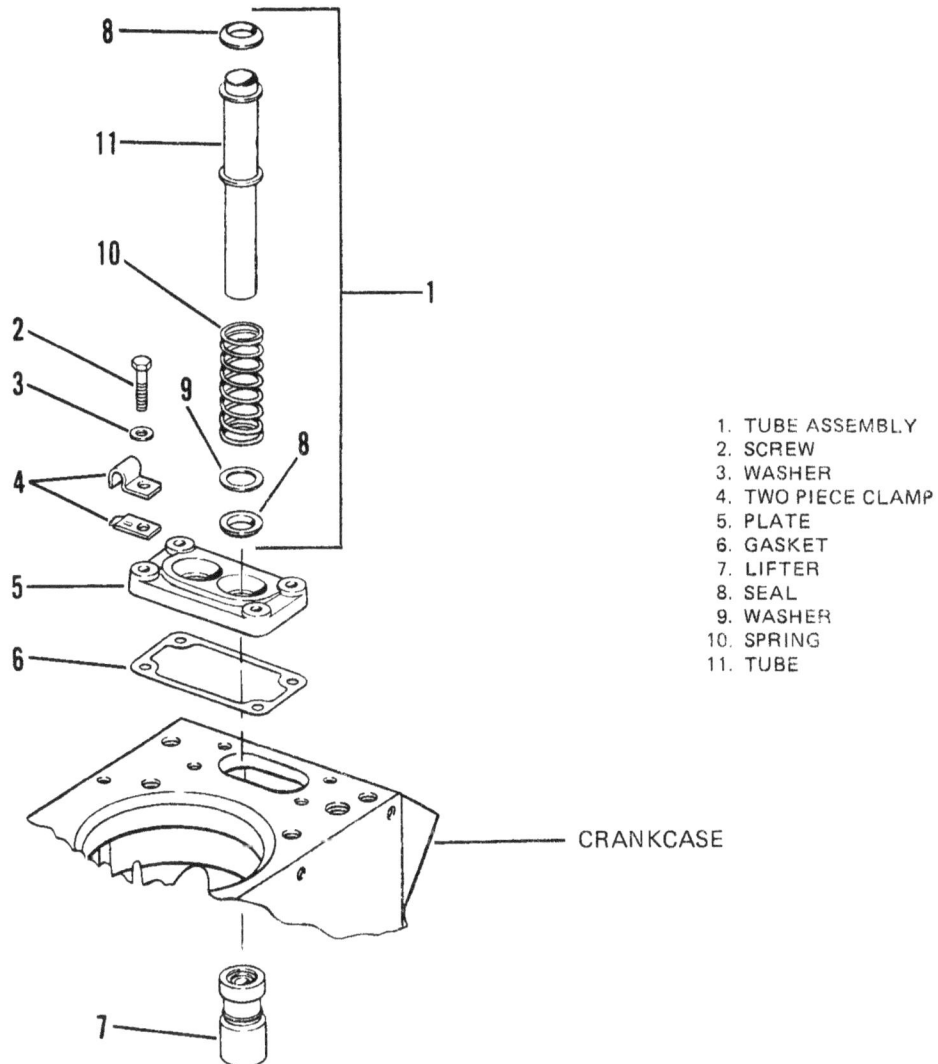

1. TUBE ASSEMBLY
2. SCREW
3. WASHER
4. TWO PIECE CLAMP
5. PLATE
6. GASKET
7. LIFTER
8. SEAL
9. WASHER
10. SPRING
11. TUBE

CRANKCASE

Figure 7-21. Push Rod Tubes and Lifters.

(3)   Remove capscrews (2) washers (3), and two-piece clamp (4).

(4)   Remove push rod plate (5) and gasket (6).

(5)   Use a magnetic pickup tool to remove lifter (7) from bore.

b.   Disassembly.

(1) Remove rubber seals (8).

(2) Remove washer (9) and spring (10) from push rod tube (11).

c. Cleaning and Inspection.

(1) Wipe parts with a clean lint-free cloth that has been slightly moistened with solvent.

(2) Inspect all parts for cracks, distortion, or other visible damage.

d. Assembly.

(1) Place spring (10) and washer (9) on push rod tube (11).

(2) Lubricate and install seals (8) on both sides of   tube.

e. Installation.

(1) Apply lifter   pre-lube (Table 2-2, item 4)   to the   lifter (7) wearing
    surfaces.

(2) Place lifter   into its bore in crankcase.

(3) Install push   rod plate (5) and gasket (6) facing cylinder   Make   sure angled edge on push
    rod plate is   head.

(4) Apply sealing compound (Table 2-2, item 7) to and install screw (2), washer (3) and two-piece clamp (4). Torque screw to 7-8 ft lbs (9.5 to 10.8 N.m).

(5) Place push rod tubes in position.

(6) Install cylinder head, refer to paragraph 7-6.

7-10.   FLYWHEEL AND ENGINE FAN.

a. Removal.
Remove frame, refer to paragraph 3-2.

(1)   Remove lifting eye and wrappers, refer to paragraph 7-6.a (1), Removal.

(2)   Remove sound shield (1, Figure 7-22) and air scroll (2) as an assembly. Remove screws (3) inserts (4), screw (5) and washer (6) through access hole in sound shield.

(3)

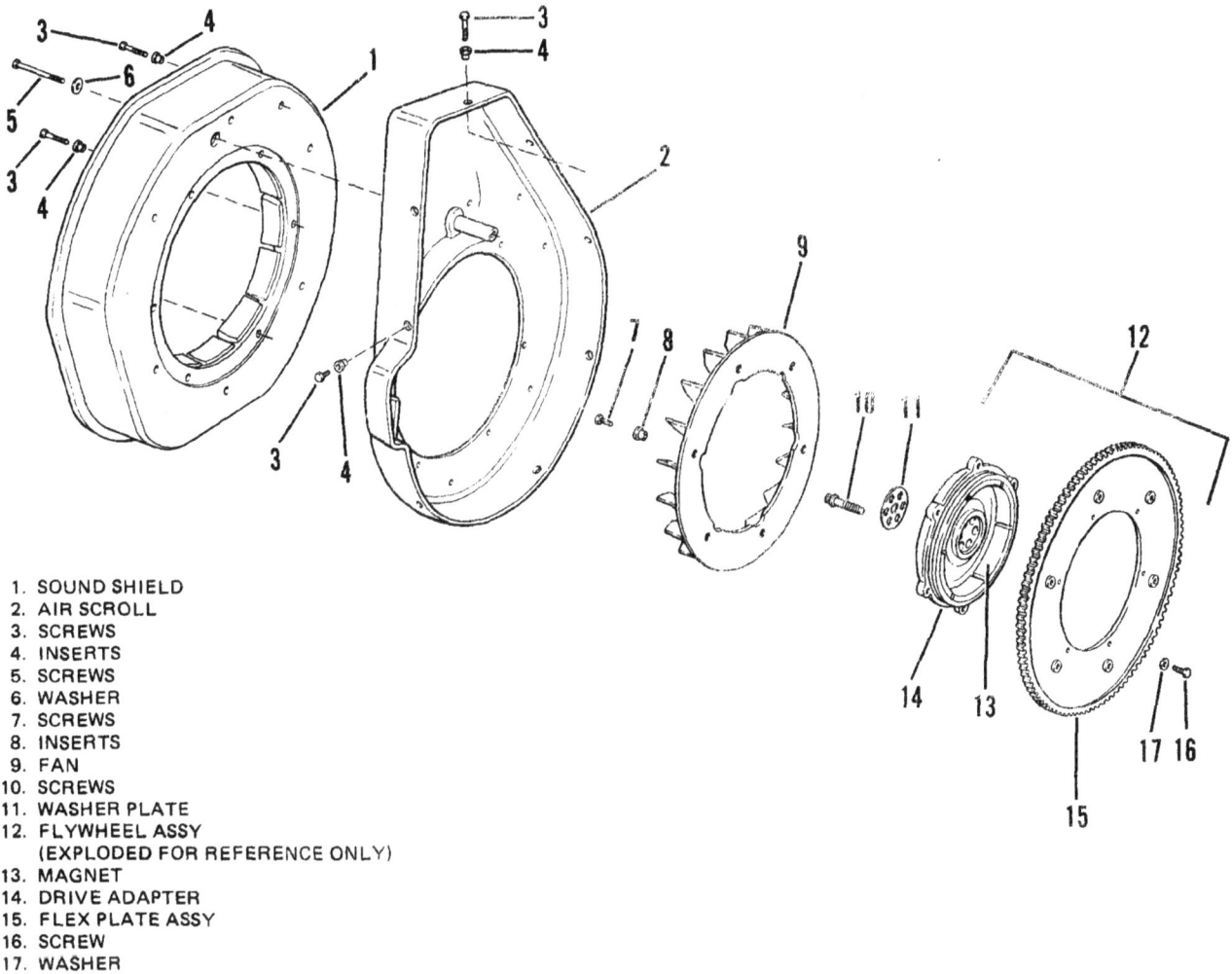

1. SOUND SHIELD
2. AIR SCROLL
3. SCREWS
4. INSERTS
5. SCREWS
6. WASHER
7. SCREWS
8. INSERTS
9. FAN
10. SCREWS
11. WASHER PLATE
12. FLYWHEEL ASSY
    (EXPLODED FOR REFERENCE ONLY)
13. MAGNET
14. DRIVE ADAPTER
15. FLEX PLATE ASSY
16. SCREW
17. WASHER

Figure 7-22. Flywheel and Engine Fan.

NOTE

It is not necessary to separate air scroll from sound shield unless parts are damaged. Loose or removed foam segments should be reattached with foam adhesive and push nuts.

Do not hold fan on flywheel to prevent engine from turning.   Doing so can

CAUTION

damage fan.

(4) Remove flywheel   and fan.

(a) Remove screws   (10), washer plate (11), and flywheel assembly (12).

(b) If necessary,   remove screws (7), inserts (8), and fan (9).

MARINE CORPS TM 05926B/06509B-34
ARMY   TM 5-6115-615-34
NAVY   NAVFAC P-8-646-34
AIR FORCE   TO 35C2-3-386-32

(1)   Clean all metallic parts in solvent and dry with compressed air.

(2)   Inspect ring gear on flywheel for cracked, worn, or otherwise damaged teeth.

Inspect fan for cracks, wear, or other damage.

(3)

Check that flywheel magnets (13) are secure on inside drive adapter (14).

(4)   If magnets are loose or damaged, flywheel assembly (12) must be replaced.

c.   Installation.

(1)   Apply sealing compound (Table 2-2, item 7) to all screws.

CAUTION

Donot hold fan on flywheel to prevent engine from turning.   Doing so
can damage fan.

(2)   Place flywheel assembly (12) on engine crankshaft.Be sure that hole in flywheel engages dowel on crankshaft.

Secure flywheel assembly with screws (10) and washer plate (11). Torque screws to 57-58 ft lbs (77-79

(3)   N.m).

Position fan (9) on flywheel assembly (12). Make sure that flat side on one of the fan slots aligns with the one

(4)   tab with a flat on the drive adapter.   Refer to Figure 7-23.

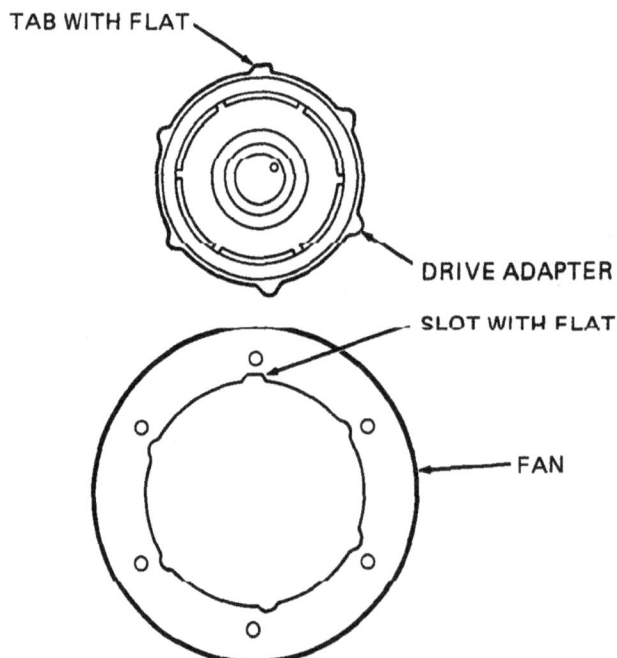

Figure 7-23.Fan and Drive Adapter Alignment Tabs.

(5) Install inserts (8, Figure 7-22) and screws (7) to secure fan. Tighten screws (7) to a torque of 5-6 ft lbs (7-8 N.m).

(6) Install air scroll (2) and sound shield (1) as an assembly.

(a) Place air scroll (2)and sound shield (1) as an assembly on engine.

NOTE

Do not apply sealing compound to air scroll mounting screws.

(b) Install screw (5) and washer (6).

(c) Install screws (3) with inserts (4).

(7) Install lifting eye and wrappers, refer to paragraph 7-6.f (8), Installation.

(8) Install frame, refer to paragraph 3-2.

7-11. BATTERY CHARGER STATOR AND AIR SCROLL BACK PLATE.

a. Testing.

(1) Use a multimeter (set to read ohms) to check the stator windings.

(2) Disconnect the stator windings at the wiring plug (1, Figure 7-24).

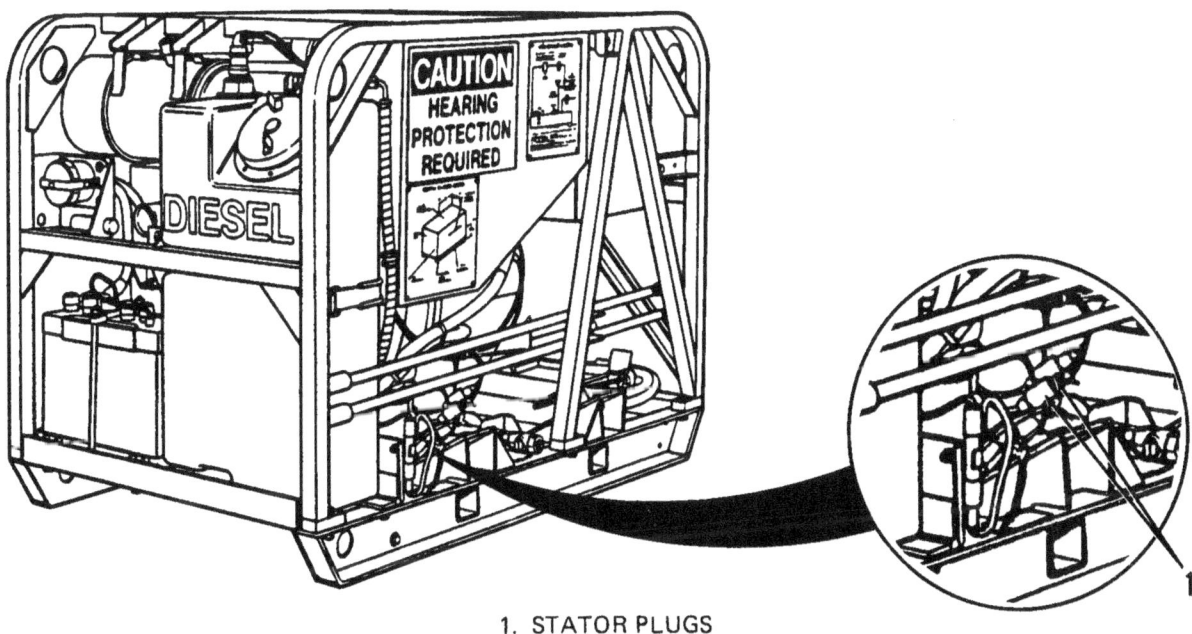

1. STATOR PLUGS

Figure 7-24.Battery Charger Stator Testing.

(3) Place the multimeter leads across the stator winding connections across plugs (1).

(4) The resistance of the stator windings should be between 0.07 and 0.12 ohms. Replace stator if resistance is not within this range.

(5) Check the resistance from each of the two stator leads and ground (engine or frame).

(6) Resistance between windings and frame or engine ground should indicate discontinuity (infinite ∞ ohms). If low resistance is indicated, stator windings are grounded and stator should be replaced.

a. Removal.

NOTE

Battery charger stator leads are partially held in position by the air scroll back plate. Stator can be removed without removing air scroll backplate if stator lead connector boots are removed first.

(1) Remove flywheel and fan, refer to paragraph 7-10.

(2) Remove screws (1, Figure 7-25).

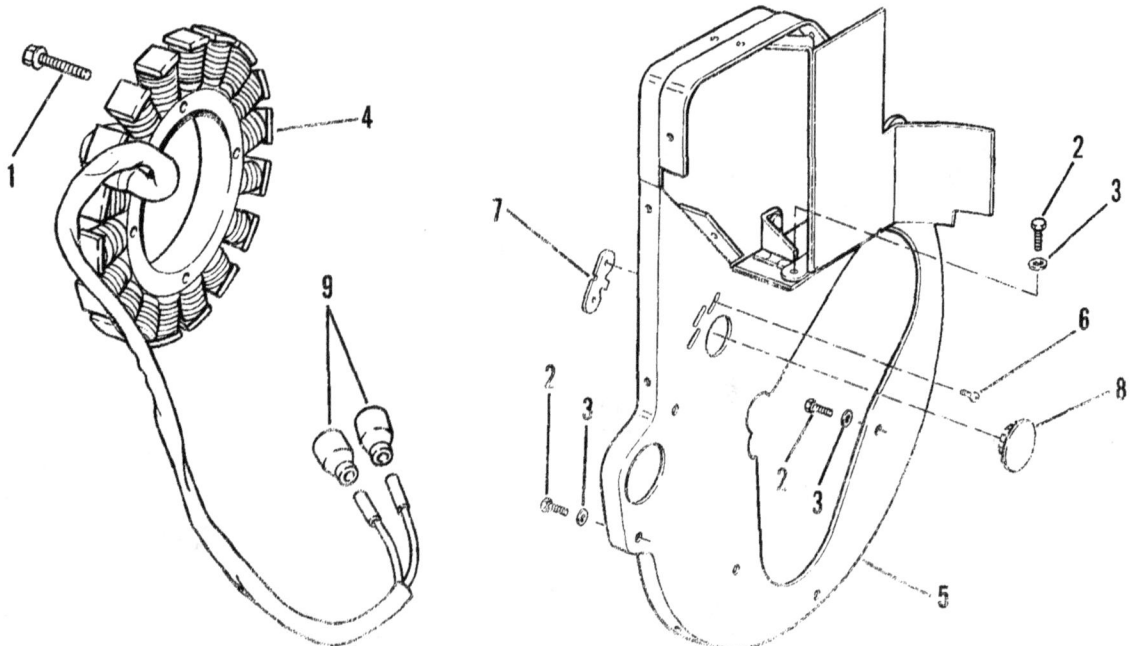

1. SCREW	4. STATOR	7. POINTER, TIMING
2. SCREW	5. AIR SCROLL BACKPLATE	8. PLUG, DUST
3. WASHER	6. SCREW	9. CONNECTOR

Figure 7-25. Battery Charger Stator.

(3) Remove screws (2) and washers (3).

(4) Disconnect stator (4) wiring at connector (9).

(5) Remove stator (4) and air scroll back plate (5).

b. _____Disassembly._____ Screws (6) and timing pointer (7) are not removed unless visibly damaged. Dust plug (8) may be removed if necessary.

c. _____Cleaning and Inspection._

(1) Wipe all parts with a clean lint-free cloth that has been slightly moistened with solvent.

(2) Inspect all parts for cracks, distortion, or other visible damage.

(3) Inspect inside of stator core for scoring or other indications of wear due to contact with drive adapter.

d. Assembly. _____If removed, loosely install timing pointer (7) and secure with screws (6).

e. Installation.

NOTE

Setting of timing pointer requires that a timing wheel be installed on the crankshaft. Timing pointer should be adjusted whenever air scroll back plate is loosened or removed.
Apply sealing compound (Table 2-2, item 7) to all screws.

Place air scroll back plate (5) in position and loosely install screws (2) and washers (3).

(1) Thread stator wiring past air scroll back plate.

(2) Place stator (4) in position and secure with screws (1).

    Tighten screws (2) securely.
(3)
    Set timing pointer as follows:
(4)
        Install flywheel, refer to paragraph 7-10.
(5) Remove cylinder head, refer to paragraph 7-6.

(6) Use 0.5 in. (12.5 mm) bar stock (two holes drilled to fit, over cylinder retaining studs), refer to Table 2-1 to retain cylinder in crankcase. Tighten bar stock lightly in a crisscross pattern.

    (a)

    (b)

    (c)

(d) Mount a dial indicator to read off of   top of piston over wrist pin.

(e) Use dial indicator to locate top dead   center of piston travel.

(f) Rotate engine to bring piston up to 0.039 in. (1 mm) before top dead
center (TDC) as indicated on dial indicator. Make a mark on the flywheel in line with the timing pointer.

(g) Bring piston up to TDC and then go to 0.039 in. (1 mm) past TDC as indicated on the dial indicator.Make a
second mark on the flywheel in line with the timing pointer.

(h) Carefully measure the distance between the first and second marks and determine the halfway point between
them.Make a third mark on the flywheel at this halfway point.

(i) Rotate the engine to put this third mark in line with the timing pointer. This is top dead center. If necessary,
adjust the location of the timing pointer so that it is directly over the TDCmark stamped on the flywheel, and
tighten screws securely.

(j) Install dust plug (8)

(7) Install lifters and push rod tubes, refer to paragraph 7-9.

7-12.   GEAR COVER AND SEAL.

a. Removal.

Remove governor, refer to TM 05926B/06509B-12/TM 5-6115-615-12/NAVFAC P-8-646-12/TO 35C2-3-386-31 manual,
paragraph 4-44.

Remove lifting eyes and wrappers, refer to paragraph 7-6 a, Removal.

(1)   Remove air scroll, sound shield, and flywheel, refer to paragraph 7-10.

(2)   Remove stator and air scroll back plate, refer to paragraph 7-11.

(3)   Remove screws (1, Figure 7-26), washers (2), cover (3) and gasket (4).

(4)   Removal of gear cover back plate (5) and gasket (6) requires removal
       of camshaft (refer to paragraph 7-16), and the oil pump (refer to paragraph 7-14).

(5)   Remove screw (7), washer (8), cover back plate (5) and gasket (6).
       b. Disassembly.Front oil seal (9) is not removed unless there is evidence of
(6)   the leakage of oil.

       c.Cleaning and Inspection.

(7)   (1) Wipe parts with a clean lint-free cloth that has been slightly moistened

       with solvent.

(2) Inspect all parts for cracks, distortion, or other visible damage.

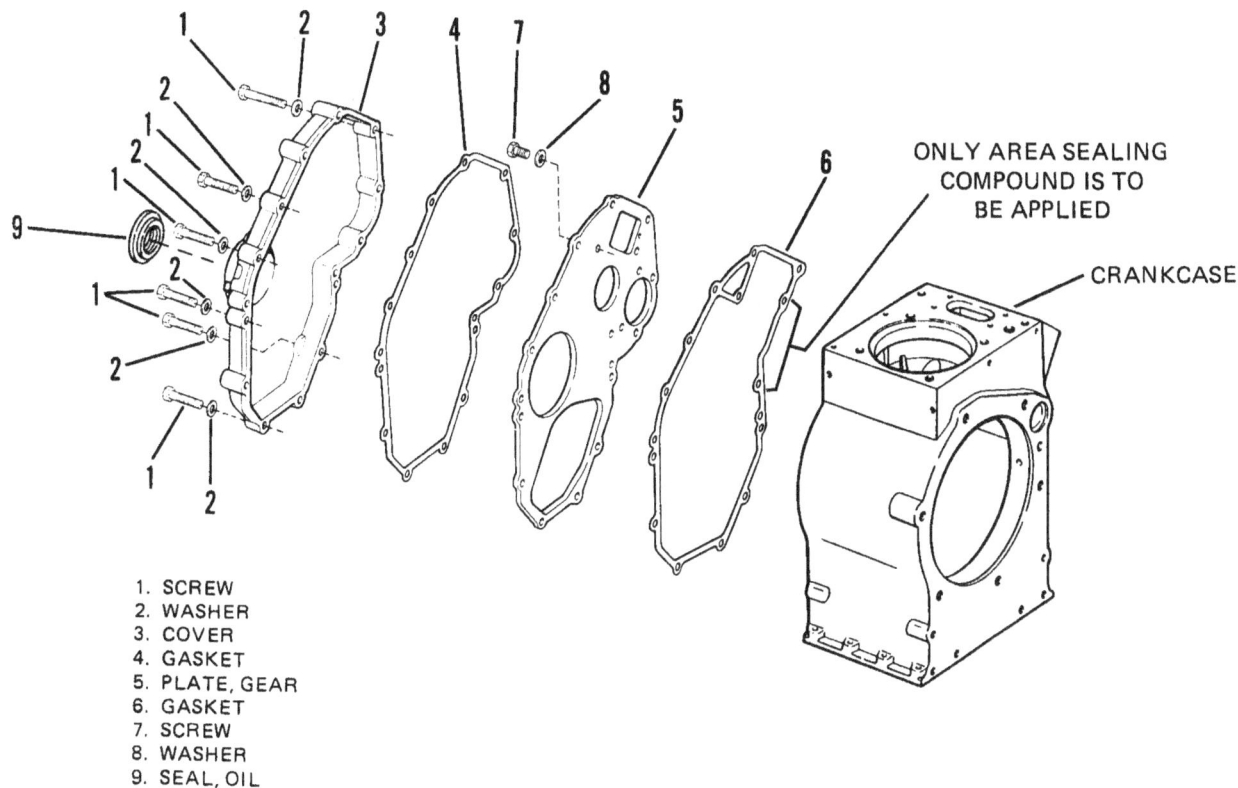

Figure 7-26.Gear Cover and Seal.

1. SCREW
2. WASHER
3. COVER
4. GASKET
5. PLATE, GEAR
6. GASKET
7. SCREW
8. WASHER
9. SEAL, OIL

d.   Assembly.

(1)   If removed, oil seal (9) can be pressed into cover (3) with a press and installation tool (refer to Table 2-1).

(2)   Seal (9) should be pressed in 0.12 to 0.17 in. (3 to 3.8 mm) from the cover face.   Be sure that oil seal lip does not run in previous oil seal lip groove.

e. Installation.

NOTE

Apply a bead of sealing compound (Table 2-2, item 10) in area shown in Figure 7-26 to both sides of gasket (6) before installation.

(1) Install cover back plate (5) with gasket (6) and secure with screw (7) and washer (8).

(2) Install camshaft, refer to paragraph 7-16.

(4) Install oil pump, refer to paragraph 7-14.

(4) Lubricate oil seal (9) with lubricating oil.

(5) Apply sealing compound (Table 2-2, item 7) to screws (1).

(6) Place gasket (4) and cover (3) in position.

(7) Install screws (1) and washers (2) in recessed bosses on cover.Bosses that are not recessed are used to mount other covers.

(8) Tighten screws in a crisscross pattern in two steps to 6.6 to 7.4 ft lbs (9-10 N.m).

(9) Install stator and air scroll back plate, refer to paragraph 7-11.

(10) Install air scroll, sound shield, and flywheel, refer to paragraph 7-10.

(11) Install lifting eyes and front and rear wrappers, refer to paragraph 7-6.f, step (9).

(12) Install governor, refer to TM 05926B/06509B-12/TM 5-6115-615-12/NAVFAC P-8-646-12/TO 35C2-3-386-31 manual, paragraph 4-44.

7-13.CYLINDER AND PISTON.

a. Removal.

(1) Remove cylinder head, refer to paragraph 7-6.

(2) Remove push rod tubes and lifters, refer to paragraph 7-9.

NOTE

Do not remove cylinder head studs (10) unless damaged.

(3) Remove cylinder (1, Figure 7-27) from piston assembly (2).

(4) Remove two circlips (3) and wrist pin (4). Remove piston(5) from connecting rod.
b. Disassembly.

(1) Remove rings (6 and 7) from piston with a ring expander.

(2) Remove oil ring (8) from piston.

(3) Remove shims (9) from cylinder.

c. Cleaning, Inspection and Repair.

(1) Wipe all parts with a clean lint-free cloth that has been slightly moistened with solvent.

(2) Ring grooves may be cleaned with a groove cleaner (refer to Table 2-1).

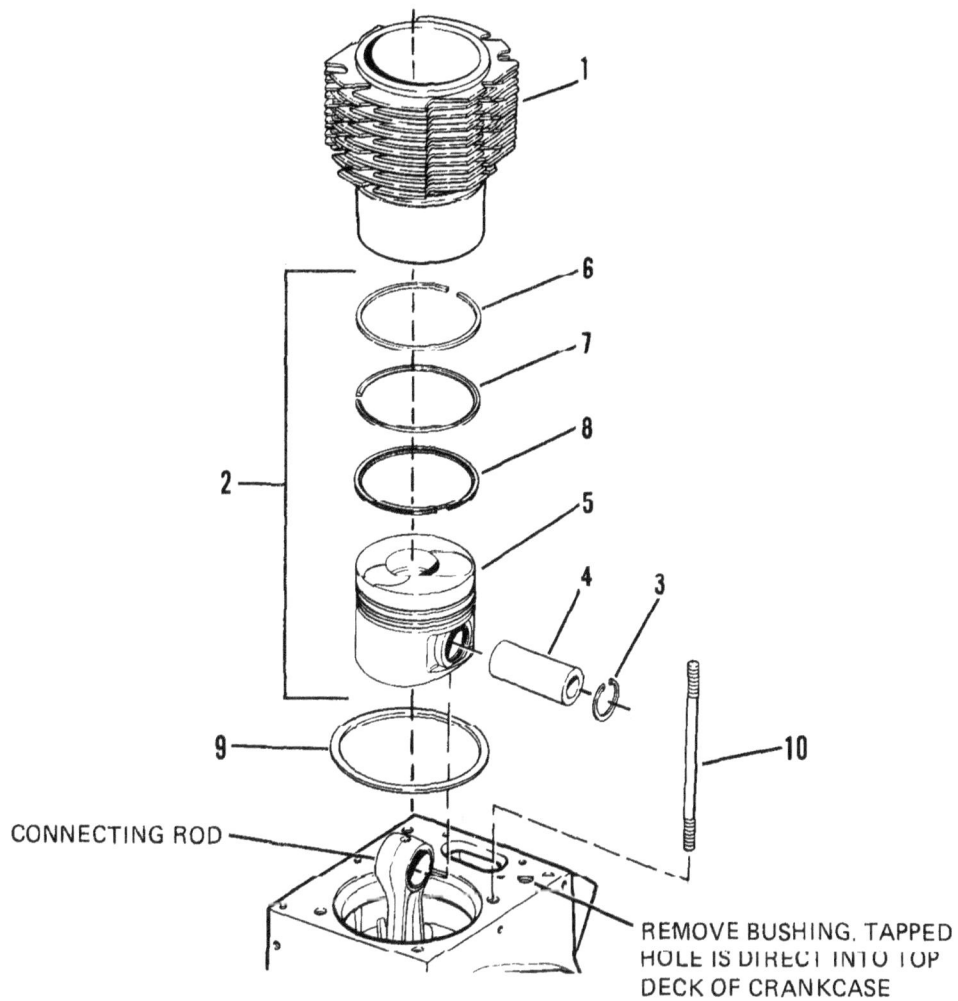

1. CYLINDER   6. PISTON RING
2. PISTON ASSY   7. PISTON RING
3. CIRCLIP   8. PISTON RING
4. WRIST PIN   9. SHIM
5. PISTON   10. STUD

CONNECTING ROD

REMOVE BUSHING, TAPPED HOLE IS DIRECT INTO TOP DECK OF CRANKCASE

(3) Inspect piston and cylinder for signs of burning, scoring, or other visible damage.

(4) Check that piston and cylinder dimensions are within specifications, refer to Table 1-1.

(5) Inspect all parts for cracks, distortion, or other visible damage.

(6) Place piston rings in cylinder.   Square rings in cylinder with piston and check that ring gapsare within specifications, refer to Table 1-1.

(7) Hone cylinder sleeve.

CAUTION

Do not place cylinder sleeve in a vise for honing.
Cylinder sleeve will be made out of round if pressure is applied to cylinder sleeve.

(a) Place the cylinder sleeve into a holding fixture (old crankcase or equivalent).

(b) Use a cylinder hone (refer to Table 2-1) and hone cylinder sleeve to achieve a 40 to 50 degree cross pattern.

(c) Clean cylinder sleeve with soap and water and apply a coat of clean engine oil.

d. Assembly.

(1) Install oil ring (8) spring in piston groove and then install oil ring. Make sure spring is seated inside oil ring.

(2) Using a ring expander, install rings (6 and 7). Markings on rings should face up.

(3) Install one circlip (3) on piston.Make sure sharp edge on circlip faces out and round edge of circlip is facing piston.

e. Installation.

(1) If removed, apply sealing compound (Table 2-2,item 5) to and install cylinder studs (10). End of studs should be 5.55 to 5.63 in. (141 to 143 mm) from crankcase face.

(2) Position ring end gaps so they are approximately 120 degrees apart. Refer to Figure 7-28.

(3) Position shims (9, Figure 7-27) on cylinder (1).

(4) Lubricate rings, piston, ring compressor and cylinder with lubricating oil.

(5) Using a ring cyl-   compressor (refer to Table 2-1), install piston (5) into
inder.

(6) Allow bottom   of piston to extend from cylinder, but keep rings retained in
cylinder.

(7) Move connecting rod to its highest point of travel.

(8) Position cylinder/piston assembly on connecting rod.Flat sides of cylinder should face front and back of engine. Markings on top of piston should face the front of engine (gear end).

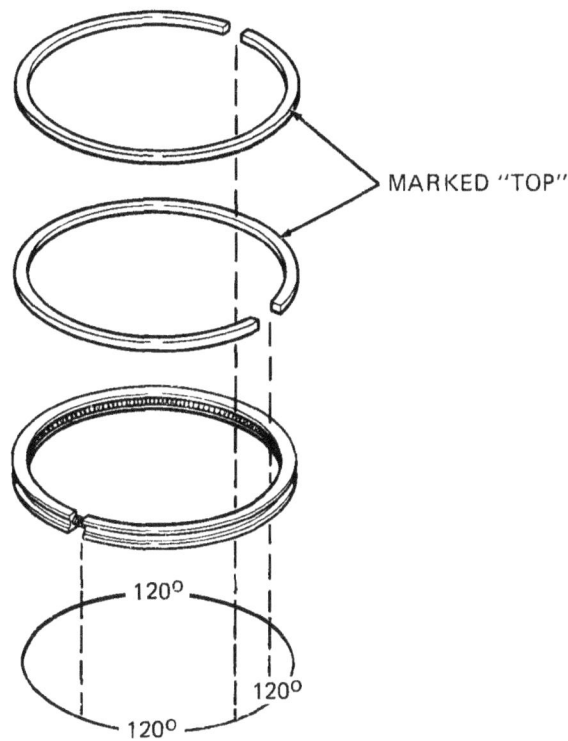

Figure 7-28.Piston Ring Positioning.

(9) Install wrist   pin (4) and circlip (3).

(10) Push cylinder   down into crankcase.

(11) Check bump clearance (piston to cylinder head clearance) as follows:

  (a)   Bring piston up near to top dead center (TDC).

  (b)   Cut two pieces of 1/16 in.solid core solder approximately 0.39 to 0.49 in. (10-12 mm) length.

       Curve me end of solder to prevent rolling.

  (c)
       Place the two pieces of solder on piston.

  (d)

  (e)   Solder should be back of positioned over wrist pin of piston at the front and
       piston.

  (f)   head and secure with cylinder head washers and nuts.

  (g)   Lubricate cylinder studs with lubricating oil to ensure proper torque.

  (h)   Snug cylinder head nuts down.Tighten nuts in a crisscross pattern in three steps 15,26, 33 ft lbs (20, 35,
       45 N.m).

  (i)   Rotate engine over to compress solder.

(j)   Remove cylinder head.

(k)   Remove and measure the thickness of the pieces of solder. Add the measurements together and divide by two to find average.

(l)   Correct bump clearance should be 0.033 to 0.037 in. (0.85 to 0.95 mm).

(m)   Add or remove shims (9) as required to bring bump clearance within specifications. Thickness of shims may be determined by the number of indented holes in the shim surface.

No holes= 0.0016 to 0.0024 in. (0.04 to 0.06 mm).

[1] One hole= 0.0055 to 0.0063 in. (0.14 to 0.16 mm).

[2] Two holes= 0.0130 to 0.0146 in. (0.33 to 0.37 mm).

[3]

(12) Recheck bump clearance, repeat step 11.

(13) Install push rod tubes and lifters, refer to paragraph 7-9.

(14) Install cylinder head, refer to paragraph 7-6.

7-14.   OIL PUMP.

a. Removal.

(1) Remove gear cover, refer to paragraph 7-12.

(2) Remove two screws (1, Figure 7-29) and washers (2).

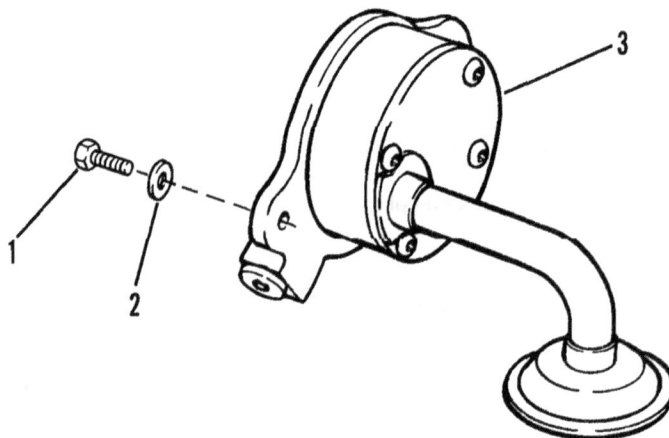

1. SCREW
2. WASHER
3. OIL PUMP

Figure 7-29. Oil Pump.

MARINE CORPS TM 05926B/06509B-34
ARMY   TM 5-6115-615-34
NAVY   NAVFAC P-8-646-34
AIR FORCE TO 35C2-3-386-32

(3) Remove oil pump assembly (3)from crankcase.   If tight, thread two M6 x 1.0
   screws into tapped holes for attachment of puller.

b. Installation.

  (1) Lubricate outside surface of pump with lubricating oil.

  (2) Position pump (3) in crankcase.Pump may need to be lightly tapped into place with a rawhide mallet.

  (3) Apply sealing compound (Table 2-2, item 7) to screws (1) and install screws (1) and washers (2).   Tighten screws to a torque of 8 ft lbs (11 N.m).

  (4) Install gear cover, refer to paragraph 7-12.

7-15. CONNECTING ROD.

a. Removal.

  (1) Remove oil pan, refer to paragraph 7-3.

  (2) Remove cylinder and piston, refer to paragraph 7-13.

  (3) Remove nuts (1, Figure 7-30).

  (4) Remove connecting rod cap (2) with bearing shell (3).

CAUTION

   Care should be taken to ensure that exposed rod bolts do not contact and damage crankshaft.

  (5) Remove connecting rod (4) with bearing shell (5).

  (6) Bolts (6) are pressed into place and are not removed unless visibly damaged.
  b. Disassembly.

  (1) Remove bearing shells (3 and 5).

  (2) Bearing (7) is not removed unless visibly damaged or worn beyond specifications. Refer to Table 1-1 for bearing specifications.

c. Cleaning and Inspection.

  (1) Wipe all parts with a clean lint-free moistened with solvent.   Inspect bearing shells for scoring or cloth that has been slightly

  (2)

(3) Lubricate bearing shells with lubricating oil.

(4) Push bearing shells (3 and 5) in connecting rod (4) and cap (2). Make sure tab on bearing shell fits into groove on connecting rod and cap.

(5) Position connecting rod cap (2) on connecting rod (4).   Make sure number printed on cap (2) is on same side as number on connecting rod (4).

(6) Lightly retain connecting rod in a soft jaw vise. Install and tighten nuts (1) to a torque of 56-63 ft lbs (76-85 N.m) in three steps 22, 37, 56 ft lbs (30, 50, 85 N.m).

(7) Measure the diameter of bearing shells (3 and 5).   If diameter is not within the limits specified in Table 1-1, replace bearing shells (3 and 5) and recheck dimensions.   If dimensions are still not within specifications, the connecting rod must be replaced.

(8) Disassemble connecting rod.

      (a) Remove nuts (1) and cap (2).

      (b) Bearing shells (3 and 5) do not need to be removed from cap (2) or connecting rod (4).

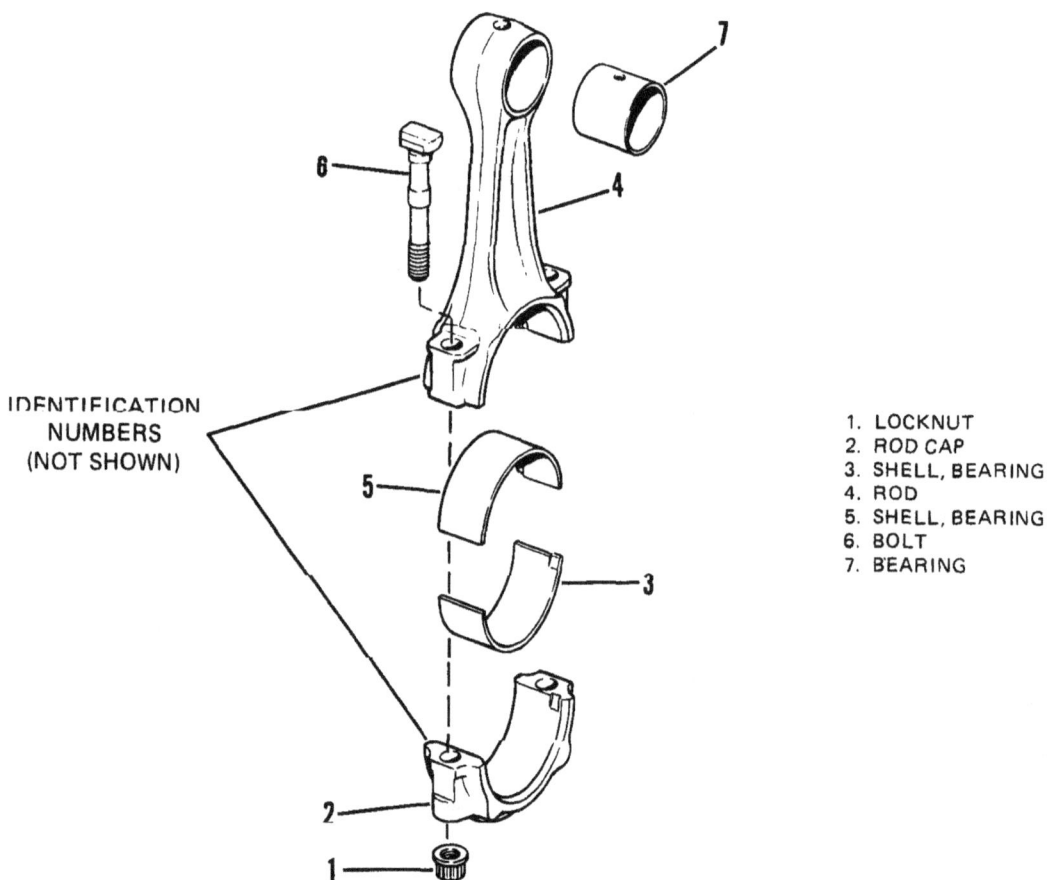

IDENTIFICATION
NUMBERS
(NOT SHOWN)

1. LOCKNUT
2. ROD CAP
3. SHELL, BEARING
4. ROD
5. SHELL, BEARING
6. BOLT
7. BEARING

Figure 7-30.Connecting Rod.

d.   Installation.

(1)   Apply a light coating of oil to bearing shells.

(2)   If removed, install bearing (7). Make sure oil holes are aligned and press bearing into place with a press and bearing

(3)   installation tool.
                                        rod is facing the camshaft side of engine.

(4)   Install connecting rod cap (2).   Make sure numbered side aligns with
        numbered side on connecting rod   (4).

(5)   Install nuts (1) and tighten to 37, 56 ft lbs (30, 50, a torque of 56-63 ft lbs in three steps   22,
        85 N.m).

(6)   Check that connecting rod moves   freely on crankshaft journal.

(7)   Install cylinder and piston, refer to paragraph 7-13.

(8)   Install oil pan, refer to paragraph 7-3.

7-16. CAMSHAFT.

a. Removal.

(1)   Remove gear cover, refer to paragraph 7-12.

(2)   Remove cylinder head, refer to paragraph 7-6.

(3)   Remove push rod tubes and lifters, refer to paragraph 7-9.

(4)   Remove fuel transfer pump, refer to TM 05926B/06509B-12/TM 5-6115-615-12/NAVFAC P-8-646-12/TO
        35C2-3-386-31 manual, paragraph 4-34.

(5)   Remove fuel injection pump, refer to paragraph 6-3.

(6)   Remove oil pump, refer to paragraph 7-14.

(7)   Remove screws (1, Figure 7-31) and washers (2).

CAUTION

        Do not nick or damage camshaft or camshaft bearings when removing or
        installing camshaft.

(8)

b. Disassembly.

(1) Camshaft drive gear (3) and governor drive gear (4) are not removed unless
    gears are visibly damaged or worn.

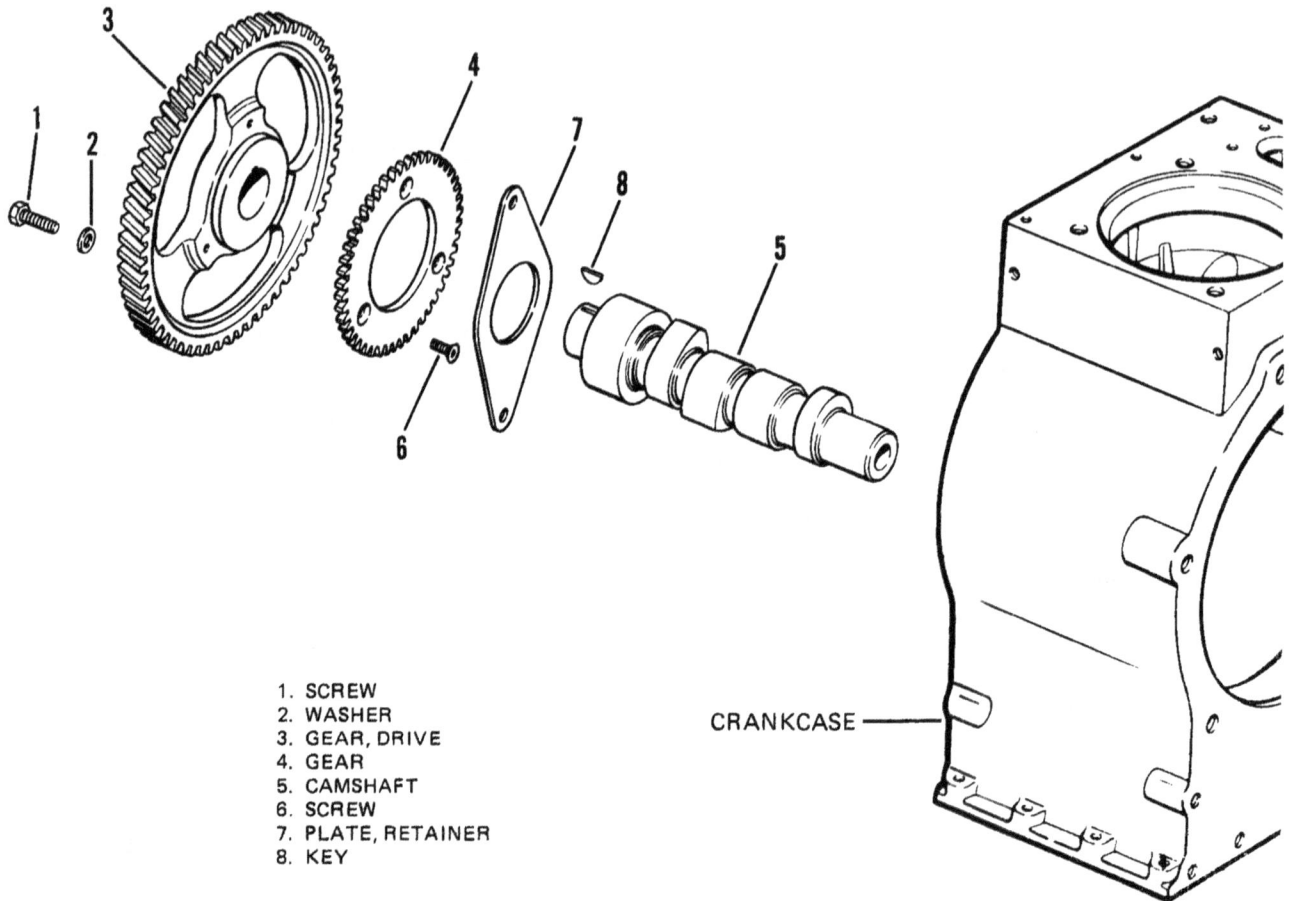

1. SCREW
2. WASHER
3. GEAR, DRIVE
4. GEAR
5. CAMSHAFT
6. SCREW
7. PLATE, RETAINER
8. KEY

CRANKCASE

(2)   If necessary, press gear (3) from camshaft (5).

(3)   Separate governor drive gear (4) from camshaft gear (3) by removing three screws (6).

      Retainer plate (7) can be removed after drive gears are removed.
(4)
      Key (8) is not removed unless key is visibly damaged.
(5)

(1)   Wipe all parts with a clean lint-free cloth that has been slightly moistened with solvent.

      Inspect gears for missing, worn, or otherwise damaged teeth.
(2)
      Inspect camshaft for excessive wear or scoring.   Refer to Table 1-1 for wear limits.
(3)
      Inspect all parts for cracks, distortion, or other visible damage.

(4)

d.   Assembly,

(1)   Attach governor drive gear (4) to camshaft drive gear (3).Apply sealing compound (Table 2-2, item 7) to screws (6) and tighten to 3.0 ft lbs
(4 N.m).

(2)   If removed install key (8).

(3)   Position retainer plate (7) on camshaft.

(4)   Press camshaft (5) into camshaft drive gear (3). Make sure retainer plate is on camshaft shoulder.

Check that retainer plate (7) moves freely on camshaft shoulder.
(5)

e. Installation.

(1)   Apply engine oil to crankcase bearing and camshaft bearing surfaces, place camshaft in crankcase, making sure camshaft does not scratch bearing surfaces.

Align timing mark (punch hole on gear tooth face) on camshaft with timing mark on crankshaft gear.
(2)
Apply sealing compound (Table 2-2, item 7) to screws (1) and install screws (1) and washers to a torque of 7-8 ft lbs (9.5-10.8 N.m).
(3)

(4)   Install   oil pump, refer to paragraph 7-14.

(5)   Install   fuel injection pump, refer to paragraph 6-4.

(6)   Install   fuel transfer pump, refer to TM 05926B/06509B-12/TM 5-6115-615-12/NAVFAC P-8-646-12/TO 35C2-3-386-31 manual, paragraph 4-34.

(7)   Install push rod tubes and lifters, refer to paragraph 7-9.

(8)   Install cylinder head, refer to paragraph 7-6.

(9)   Install gear   cover, refer to paragraph 7-12.

7-17.   CRANKSHAFT AND GENERATOR ADAPTER.

a. Removal.

(1) Remove   connecting rod,   refer to paragraph 7-15.

(2) Remove   camshaft, refer   to paragraph 7-16.

(3) Remove   oil pump, refer   to paragraph 7-14.

NOTE

Use holding bar to hold crankshaft crankshaft drive gear (4).   stable when removing drive

(4)   Using crankshaft gear puller (refer to Table 2-1), remove crankshaft gear (4).

(5)   Remove screws (1, Figure 7-32) and washers (2) that secure generator adapter (3) to crankcase.

(6)   Remove crankshaft (5).

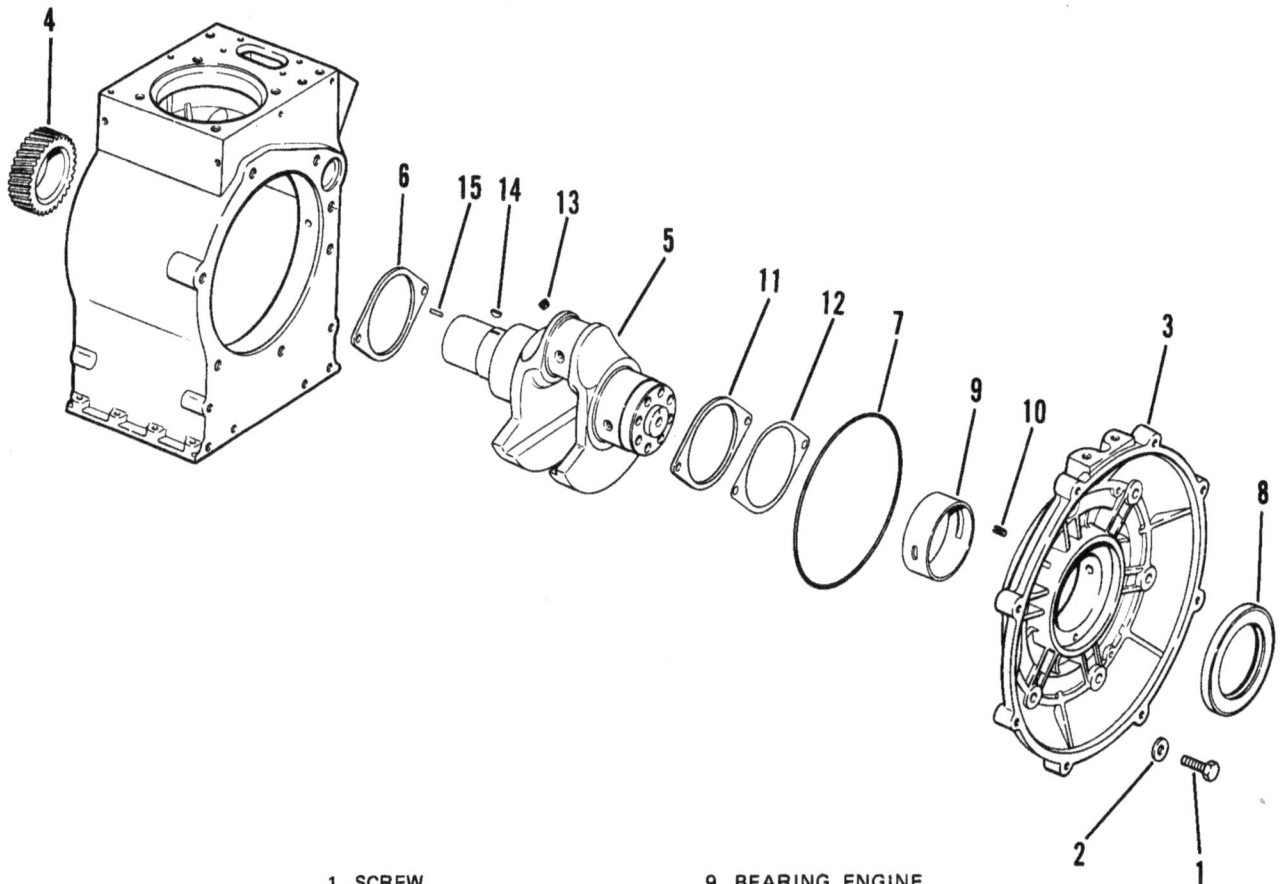

1. SCREW	9. BEARING, ENGINE
2. WASHER	10. PIN
3. GENERATOR ADAPTER	11. BEARING, THRUST
4. GEAR	12. SHIM
5. CRANKSHAFT	13. PLUG
6. BEARING, THRUST	14. KEY
7. SEAL	15. PIN
8. SEAL, OIL	

MARINE CORPS TM 05926B/06509B-34
ARMY   TM 5-6115-615-34
NAVY   NAVFAC P-8-646-34

b.   Disassembly.

    (1) Disassemble generator adapter.

       (a)   Remove thrust bearing (11) and shims (12).

       (b)   Remove seal (7) from generator adapter.

       (c)   Oil seal (8) is not removed unless visibly damaged, worn, or leaking oil.   Seal is pressed in
            and must be discarded if removed.   Removal of seal is
            accomplished with a seal puller.

       (d)   Engine bearing (9) is not removed unless visibly damaged or worn beyond specifications (refer to Table 1-1).
            Bearing is removed with a puller.

       (e)   Thrust bearing locating pins (10)are not removed unless visibly damaged. Pins (10)are pressed into
            generator adapter.

    (2) Disassemble crankshaft.Oil plug (13), woodruff key (14) and locating pin
        (15) are not removed unless they are visibly damaged, or thorough cleaning is required.

c. Cleaningand Inspection.
Wipe all parts with a clean lint-free cloth that has been slightly moistened with solvent.

    (1)   Inspect crankshaft gear for missing, worn,or otherwise damaged teeth.

        Inspect all parts for cracks, distortion, or other visible damage.
    (2)

        Check for correct crankshaft journal tolerances with a micrometer. Refer to Table 1-1 for specifications.
    (3)   d.   Assembly.

    (4)   (1) Assemble generator adapter.

     (a) If removed, install engine bearing (9).

      1 Lubricate bearing inside and out with grease.

      2 Position bearing so that oil hole in bearing is alined with generator

        adapter oil feed hole and oil groove in bearing is in top half of
        bearing.

      3 Support housing around bearing bore.   Press bearing (9) into place using a press and
        bearing installation tool (refer to Table 2-l).

      4 Bearing should be flush to or 0.02 in.(0.5 mm) below face of generator adapter.

     (c) Support flywheel housing and press oil seal (8) into generator adapter
        with a press and installation tool.

     (d) Place seal (7) on generator adapter.

MARINE CORPS TM 05926B/06509B-34
ARMY   TM 5-6115-615-34
NAVY   NAVFAC P-8-646-34
AIR FORCE   TO 35C2-3-386-32

(2) Assemble crankshaft.

    (a)Apply sealing compound (Table 2-2, item 5) to oil plug (13).

    (b) Install oil plug (13) to a torque of 6.0 ft lbs (8 N.m).

    (c) If removed, tap woodruff key (14) and locating pin (15) into place with a hammer.

e.   <u>Adjust End Play.</u>

(1)   Apply a light coat of grease to thrust bearing (6) and lightly oil crankshaft journals.

(2)   Place thrust bearing (6) on locating pins inside crankcase.Grooved side of thrust bearing must face towards crankshaft journal surface.

(3)   Slide crankshaft (5) into crankcase.Be careful that journals and bearings are not nicked or scratched. Check that thrust bearing (6) is in place.

(4)   Apply a light coating of grease to thrust bearing (11). Place thrust bearing (11) on generator adapter (3). Grooved side of bearing should face crankshaft journal.

(5)   Be sure that seal (7) is in place and install generator adapter to crankcase.   Light tapping of generator   thrust adapter with a rawhide mallet necessary.   Generator adapter should rotate freely when installed.

(6)   Tighten screws (1) in a crisscross pattern to 33 ft lbs (45 N.m).

                                                           may be

(7)   Attach a dial indicator to crankshaft end.

(8)   Move crankshaft back and forth to measure end play.Record end play. End play should be 0.008 to 0.014 in. (0.20 to 0.35 mm).

(9)   Place the required amount of shims (12) behind thrust bearing (11) on generator adapter to bring end play to 0.008 to 0.014 in. (0.20 to 0.35 mm).

(10)   Thickness of shims is determined by number of small indented holes on the shim surface.

(11)   Noholes: 0.0035 to 0.0043 in. (0.09 to 0.11 mm)

            One hole: 0.0070 to 0.0087 in. (0.18 to 0.22 mm)

f. <u>Installation.</u>

    (1) Slide crankshaft (5) into crankcase.Be careful that journals and bearings are not nicked or scratched. Check that thrust bearing (11) is in place.

(2) Be sure that seal (7) is in place and install generator adapter with thrust bearing (11) to crankcase.Light tapping of the generator adapter with a rawhide mallet may be necessary. Generator adapter should rotate freely when installed.

(3) Apply sealing compound tighten screws (1) with crisscross pattern. (Table 2-2, item 7) to screws (1), install and washers (2) to a torque of 33 ft lbs (45 N.m) in a

(4) Recheck end play, refer to step e (8 through 10).

WARNING

Hot oil and heated burn crankshaft drive gear can severely of skin.Wear insulated gloves and protective clothing to protect against burns.

(5) Heat crankshaft drive gear (4) in a hot oil tank to 350°F (166°C) (approximately twenty minutes).

CAUTION

When assembling gear (4) on crankshaft (5) make sure tapped holes are facing up.

(6) While crankshaft drive gear (4) is still hot, place gear on crankshaft and tap into place. If drive gear does not slide freely onto crankshaft, repeat step 6.

Allow crankshaft drive gear to cool for several minutes.

(7) Install oil pump, refer to paragraph 7-14.

(8) Install camshaft, refer to paragraph 7-16.

(9) Install connecting rod, refer to paragraph 7-15.

(10)

7-18. OIL FILTER ADAPTER.

a. Removal.

CAUTION

Do not apply air pressure to the oil drain process. crankcase to speed the can force the oil Air pressure sealsout of the crankcase.

(1) Drain the engine of engine oil, refer 12/NAVFAC P-8-646-12/TO 35C2-3-386-31 to TM 05926B/06509B-12/TM 5-6115-615-manual, paragraph 4-7.

(2) Remove oil filter (1, Figure 7-33).

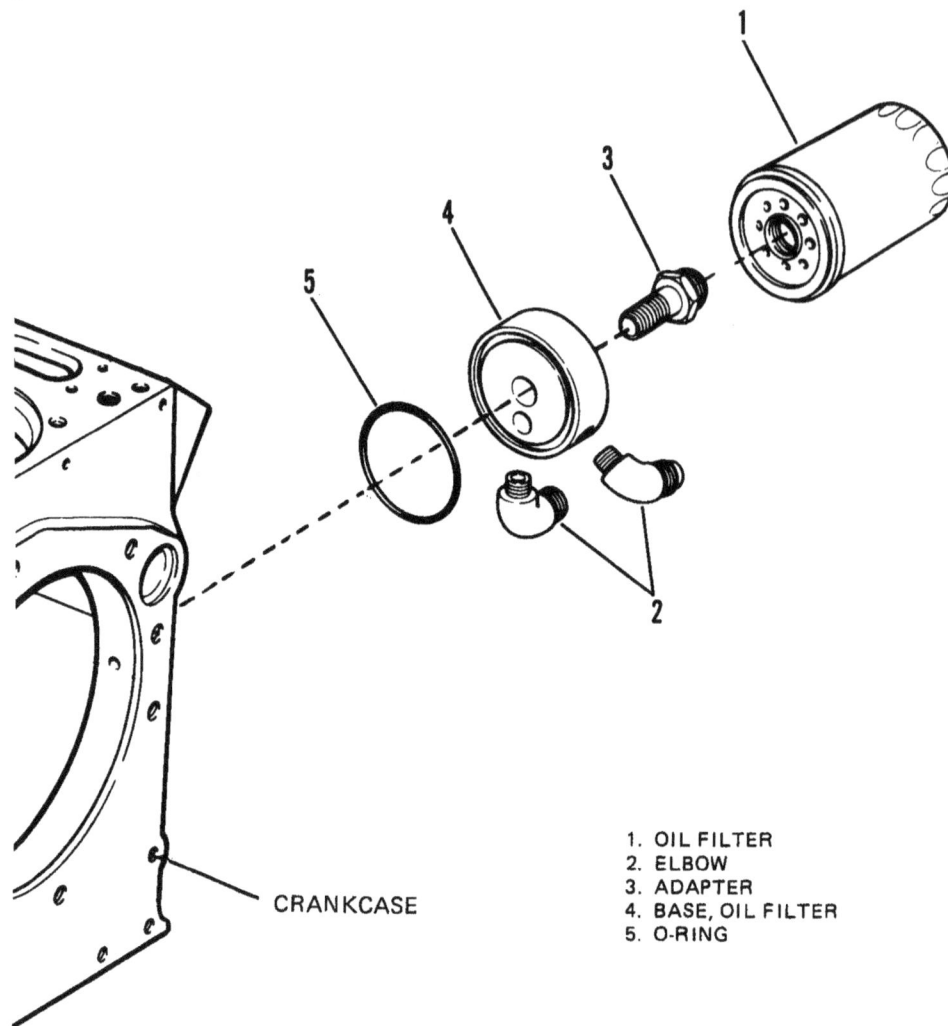

1. OIL FILTER
2. ELBOW
3. ADAPTER
4. BASE, OIL FILTER
5. O-RING

CRANKCASE

Figure 7-33. Oil Filter Adapter.

(3) Disconnect oil cooler lines from elbows (2).

(4) Unscrew threaded adapter (3) from crankcase.

(5) Remove filter base (4) and O-ring (5). Discard O-ring (5).

b. Cleaning and Inspection.

(1)   Wipe parts with a clean lint-free cloth that has been slightly moistened with solvent.

       Inspect threaded adapter and oil filter base for damaged threads, cracks, or other visible damage.
(2)

c. Installation.

NOTE

Apply Loctite (Table 2-2, item 5) to threaded adapter (3) before installation.

(1) Secure oil filter base (4) and new O-ring (5) to crankcase with threaded adapter (3). Fittings should be installed at 6 o-clock position (see Table 1-2 for torque values).

(2) Reconnect oil cooler lines to elbows (2).

(3) Lightly coat the sealing gasket of oil filter (1) with oil. Install oil filter (1).

7-19. CRANKCASE.

a. Removal.

(1) Remove crankshaft, refer to paragraph 7-17.

(2) Remove oil filter and adapter, refer to paragraph 7-18.

(3) Oil fill tube (1, Figure 7-34) is pressed into crankcase and is not removed unless it is visibly damaged.

    (a) To remove damaged oil fill tube, use a hammer and chisel to bend tube in where it enters the crankcase.

    (b) Remove and discard oil fill tube.

b. Disassembly.

(1) Bearings are not removed unless visibly damaged or worn beyond specifications.

    (a) Front main engine bearing (2).

    (b) Front camshaft bearing (3).

    (c) Governor actuator bearing (4) and seal (5).

    (d) Rear camshaft bearing (6); requires the removal of welch plug (7).

(2) Oil transfer tube (8) is not removed unless visibly damaged. Removal requires that cap plugs (9) be removed.

(3) If the crankcase oil galleries are to be cleaned, the removal of all cap plugs and pipe plugs is required. Note and record locations for reassembly.

(4)   Locating pins (10) are not removed unless visibly damaged.

(5)   Remove screws (11) lockwashers (12), washers (13) and feet (14).

(6)   Remove cup plugs (15) and pipe plugs (16).

(7)   Rear starter mounting bracket (19) is not removed unless visibly damaged.

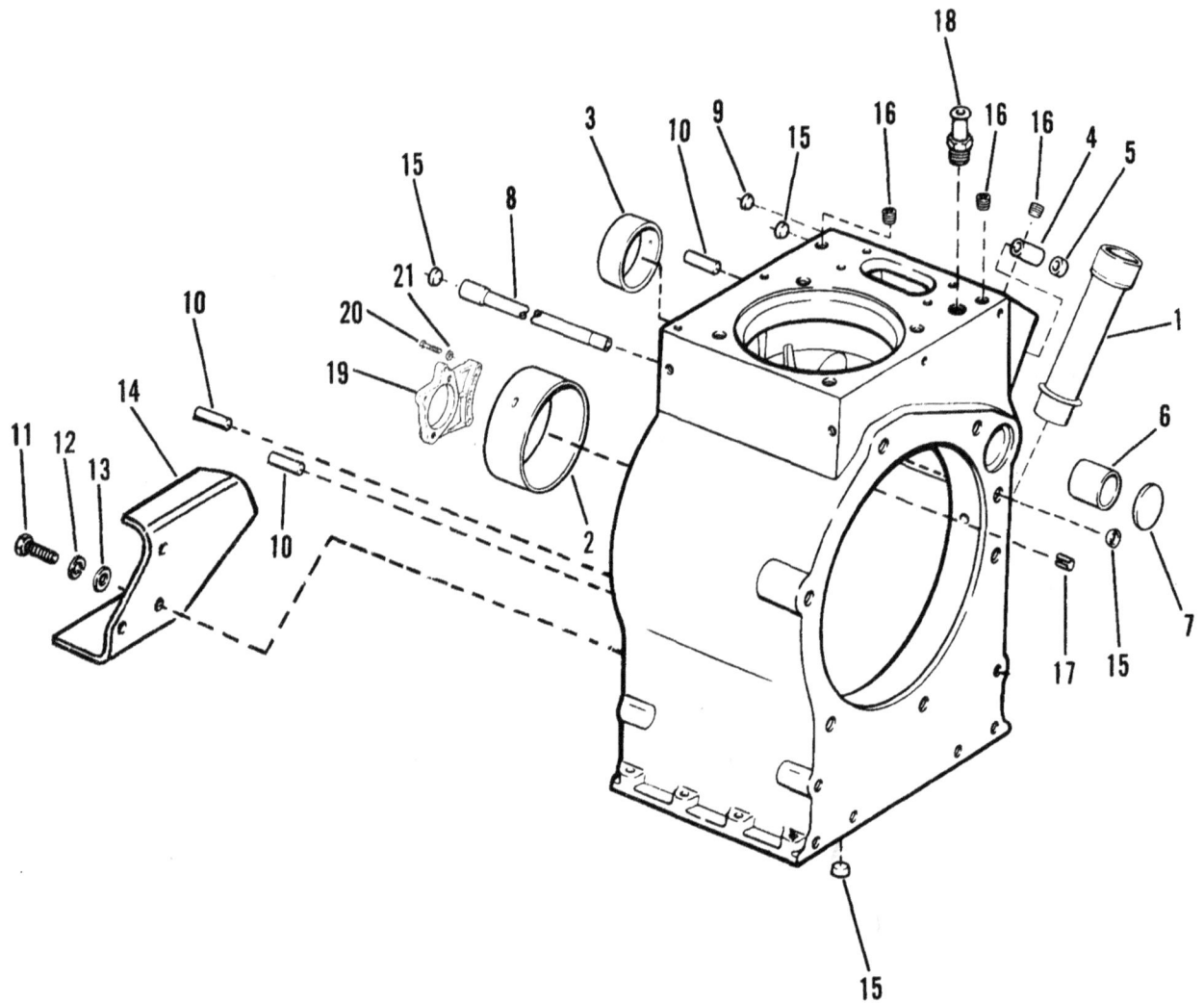

1. TUBE, OIL FILL	8. TUBE, TRANSFER	15. PLUG, CAP
2. BEARING, ENGINE	9. PLUG, CAP	16. PLUG, PIPE
3. BEARING, CAMSHAFT	10. PIN	17. PIN
4. BEARING, ACTUATOR	11. SCREW	18. BREATHER HOSE FITTING
5. SEAL	12. LOCKWASHER	19. BRACKET, STARTER, REAR
6. BEARING, CAMSHAFT	13. WASHER	20. SCREW
7. PLUG, WELCH	14. FOOT	21. WASHER

Figure 7-34. Crankcase.

MARINE CORPS TM 05926B/06509B-34
ARMY TM 5-6115-615-34
NAVY   NAVFAC P-8-646-34
AIR FORCE   TO 35C2-3-386-32

c.   Cleaning and Inspection.

(1)   General cleaning of the crankcase may be accomplished by wiping surfaces with a clean lint-free cloth.

NOTE

Do not remove governor actuator shaft bearing unless worn.   Refer to Table 1-1.

(2)   Cleaning of oil galleries requires the crankcase to be hot tanked in a cleaning solvent. Hot tanking the crankcase requires the removal of all bearings, cup plugs, and pipe plugs.

(3)   Examine bearings for excessive wear or scoring.   Check that bearing dimensions are within specifications.
Test the crankcase for cracks by dye penetration testing or magnetic particle inspection.

(4)   d.   Assembly.

(1) Install main engine bearing (2).

    (a) Lubricate bearing (2) inside and out with grease.

    (b) Aline oil holes in bearing with oil holes in crankcase, oil groove up.

    (c) Press bearing into place with a press and bearing installation tool.

(2) Install rear camshaft bearing (6).
Lubricate bearing (6) inside and out with grease.

    (a)   Bearing (6) is installed from the inside of crankcase.   Inside bore has a lead in chamfer for bearing installation.

    (b)   Aline oil hole in bearing with oil hole in bottom of crankcase bore.
        Seam on bearing should be on the oil filter side of crankcase.

    (c)
(3) Install front camshaft bearing (3).

    (a) Lubricate bearing (3) inside and out with grease.

    (b) Position bearing (3) so that oil holes are alined and that bearing seam is in the 11 o'clock position on crankcase front.

    (c) Press bearing into crankcase with a press and bearing installation tool.

(4) Install governor actuator bearing (4) and oil seal (5).

    (a) Lubricate bearing (4) and oil seal (5) inside and out with lubricating oil.

    (b) Pressbearing (4) in crankcase to a depth of 0.18 to 0.19 in. (4.50 to 4.75 mm) below crankcase surface.

    (c) Finish ream bearing (4) inside diameter to 0.311 - 0.312 in. (7.900 - 7.925 mm).

    (d) Press seal (5) into crankcase to a depth of 0.01 to 0.02 in. (0.25 to 0.50 mm) below crankcase surface.

(5) Install welch plug (7).
Apply a small amount of sealing compound (Table 2-2, item 11) to welch plug.

    (a) Position welch plug on crankcase and flatten plug with a large round drift and a hammer.

    Make sure welch plug is below the crankcase surface.

    (b) Install cup plugs (15). There are four 12 mm plugs and one 9 mm cup plug. All cup plugs are coated with a small amount of sealing compound (Table 2-2, item 11) and installed to a depth of 0.11 to 0.12 in. (2.7 to 3.2 mm) below surface of crankcase.

    (c) Install four 1/8 in. pipe plugs (16).Pipe plugs are coated with teflon sealing compound (Table 2-2,item 11) and installed to a torque of 4.4 ft lbs. (6 N.m).

If removed, install locating pins (10) and pin (17).

(7) Apply sealing compound (Table 2-2, item 7) to six screws (11).

Install two feet (14) and secure with six screws (11), lockwashers (12), and washers (13).

(8) If removed, apply Loctite (Table 2-2, item 11) and install screws (20), washers (21) and bracket (19).

(9) Installation.

(10)

(11)

e.

(1) Install oil filter and base, refer to paragraph 7-18.

(2) Install crankshaft and generator adapter, refer to paragraph 7-17.

7-20. OIL COOLER.

a. Removal.

CAUTION

Do not apply air pressure to the crankcase to speed
the oil drain process. Air pressure can force the oil seals out of the crankcase.

NOTE

To expedite the oil draining process, block unit opposite drain valve.

Drain engine oil, refer to TM 05926B/06509B-12/TM 5-6115-615-12/NAVFAC P-8-646-12/TO 35C2-3-386-31 manual, paragraph 3-30.

(1)

Remove battery, refer to TM 05926B/06509B-12/TM 5-6115-615-12/NAVFAC P-8-646-12/TO 35C2-3-386-31 manual, paragraph 4-22.

(3)

Disconnect hoses (1 and 2, Figure 7-35) from oil filter adapter elbows.

(3)

Remove screw (4) and clamp (5).

(4)

Remove screw (6), clamp (7), and remove oil cooler (8) from cover (9).

(5)

Remove hoses (1 and 2) and sleeves (3) from oil cooler (8), taking care not to damage the oil cooler.

(6)

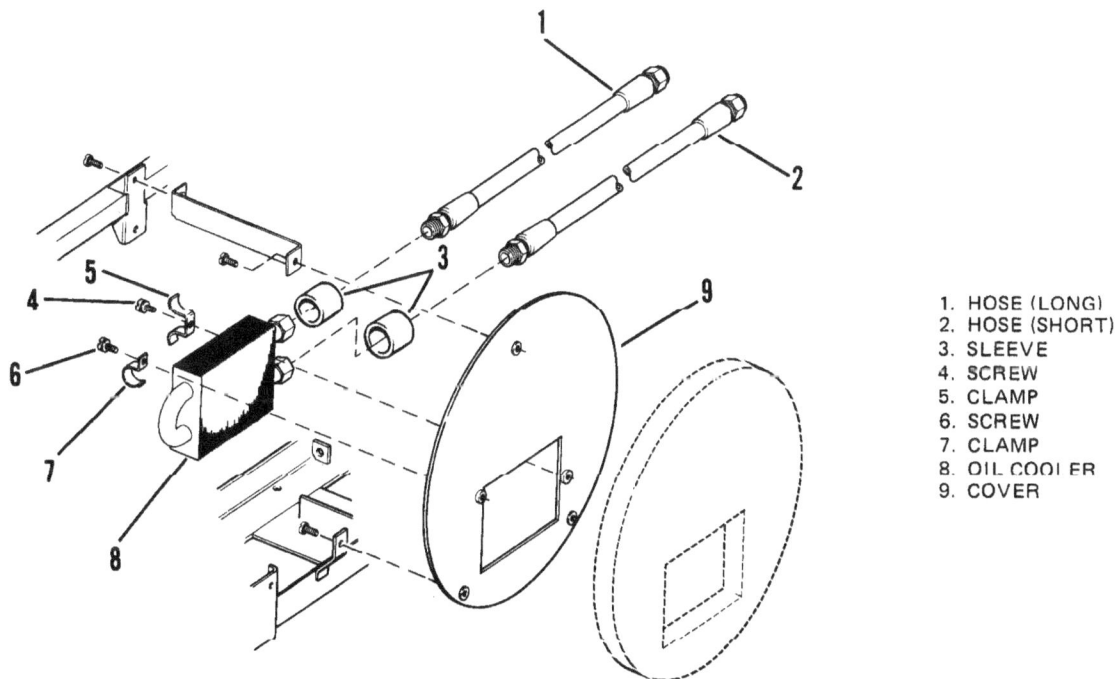

1. HOSE (LONG)
2. HOSE (SHORT)
3. SLEEVE
4. SCREW
5. CLAMP
6. SCREW
7. CLAMP
8. OIL COOLER
9. COVER

Figure 7-35. Engine Oil Cooler.

b. Repair.   Bent cooling fins may be repaired by carefully bending fins.

c. Installation.

NOTE

Be sure that the short hose is installed on top fitting of oil cooler.

(1)   Carefully install hoses (1 and 2, Figure 7-35) and sleeves (3) onto oil cooler (8).

Secure oil cooler (8) to cover (9) with clamp (7) and screw (6).

(2)   Install clamp (5) and screw (4).

(3)   Connect hoses (1 and 2) to oil filter adapter elbows.

(4)   Install battery, refer to TM 05926B/06509B-12/TM 5-6115-615-12/NAVFAC P-8-646-12/TO 35C2-3-386-31 manual, paragraph 4-22.

(5)
Fill engine with the proper amount and type of engine oil. Refer to TM 05926B/06509B-12/TM 5-6115-615-12/NAVFAC P-8-646-12/TO 35C2-3-386-31

(6)   manual, paragraph 3-30.

CHAPTER **8**

MAINTENANCE OF   ENGINE CONTROLS AND INSTRU-

8-1.GENERAL. Theengine control circuit board contains five separate circuits and the ammeter sealing resistors.

a.Field Flashing Circuit._____The field flashing circuit provides a generator
flashing current during engine starting.The flashing circuit removes the flashing current once the engine is at its normal operating speed.   Flashing of the generator field windings
ensures that the generator output will buildup properly.

b.   Glow Plug Circuit._____The glow plug circuit provides current to the engine glow plug during engine starting.
The circuit disconnects the glow plug current when the engine reaches its normal operating speed.

c.Start Disconnect Circuit.The start disconnect circuit senses the battery charging stator output.When the stator output
reaches a "breakover" value the field flashing and glow plug circuits are disconnected.

d.   Fuel Pump Circuit._____The fuel pump circuit is switched on and off by the fuel level switch in the fuel tank.
When the fuel level falls below a certain point, the fuel pump circuit activates the fuel pump. When the fuel level rises to a certain point, the fuel pump circuit switches off the fuel pump.

e.Low Fuel Cutoff Circuit._____When the fuel level in the fuel tank falls below a certain point, the low fuel
cutoff circuit will de-energize the fuel shutdown solenoid. When the fuel shutdown solenoid is de-energized, the fuel supply is
cutoff to the fuel injection pumps.

f. Circuit Protection.The circuit board uses diodes to protect against damage caused by the reversal of battery cables and voltage spikes.

8-2.   ENGINE CONTROL CIRCUIT BOARD.
Repair of the engine control circuit board is limited to the replacement of defective relays K1, K2, K3, or K4. If replacement of
relays does not correct the circuit board defect, the engine control circuit board must be replaced.

Using a multimeter and an appropriate power source, locate the defective relay on the printed circuit board.

Remove screws securing relay.

a.

b.

c.

d.   Remove the defective relay from   the relay socket.

e.   Replace new relay and screws.

f.   Test circuit board, refer to TM 646-12/TO   05926B/0650913-12/TM 5-6115-615-12/NAVFAC P-8-paragraph 4-73.
35C2-3-386-31 manual,

MARINE CORPS TM 05926B/06509B-34
ARMY   TM 5-6115-615-34
NAVY   NAVFAC P-8-646-34
AIR FORCE   TO 35C2-3-386-32

CHAPTER 9

MAINTENANCE OF GENERATOR CONTROLS AND INSTRUMENTS.

9-1. GENERAL. This chapter covers the testing and maintenance of the frequency transducer, frequency meter (60/400 Hz Sets), and control box assembly. The frequency converter on the 28 VDC generator set is used as a sensing unit for the tachometer which indicates the revolutions per minute (rpm) of the generator set. The frequency transducer on the 60 and 400 Hz generators is used as a for the frequency meter.

sensing unit

9-2. CONTROL BOX ASSEMBLY.

     a. Removal.   Refer to TM 05926B/06509B-12/TM 5-6115-615-12/NAVFAC
         35C2-3-386-31 manual, paragraph 4-58.

                                                            P-8-646-12/TO

     b. Repair. Repair of the control box assembly is accomplished by

                                                              the

     replacement of defective components. Refer to individual components in this manual and in the operator/organizational manual for procedures describing their removal.   Repair of the voltage regulator is described in paragraph 5-9 and repair of the engine printed circuit board is described in paragraph 8-2.

9-3. FREQUENCY METER. MEP-016B/MEP-021B (60/400 Hz) Sets Only.

     a. Testing.

       (1)   Connect a frequency meter which is known to be good across the frequency meter connections. Connections must be in parallel with meter being tested.

             Start and run generator set.
       (2)

       (3)   Both frequency meters should indicate the same frequency. being tested as required.   Adjust meter

             If adjustment does not bring frequency meter to frequency
       (4)   on test meter,
             replace frequency meter.

     b. Replacement. If meter is defective, follow replacement procedures given in the TM 05926B/06509B-12/TM 5-6115-615-12/NAVFAC P-8-646-12/TO 35C2-3-386-31 manual, paragraph 4-66.

9-4. FREQUENCY TRANSDUCER. a.   Testing

(On generator set).

     (1) Stop the generator set.

CAUTION

     When testing frequency transducer (Figure 9-1), a load resistance of 1100 ± 27 ohms or a frequency meter must be connected across terminals - and +.

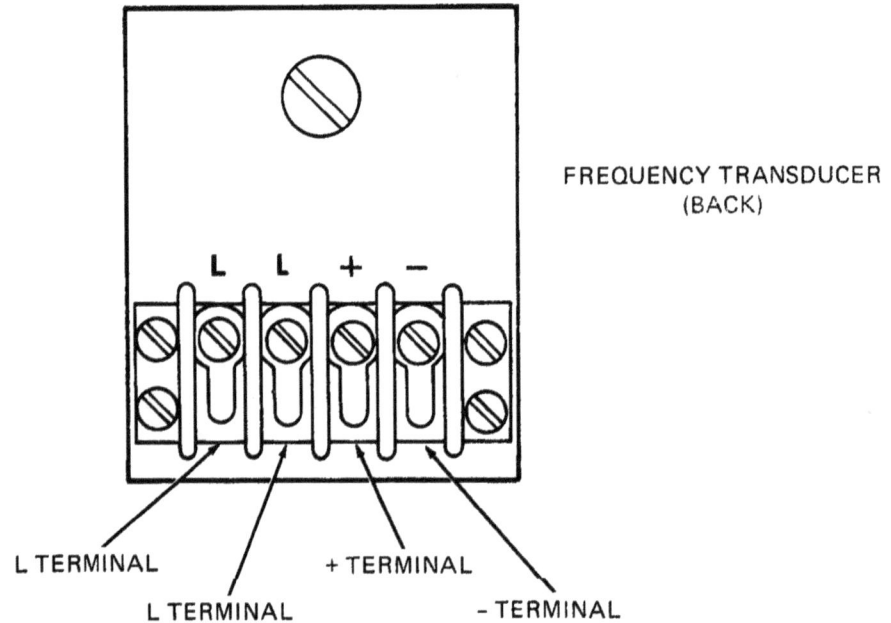

FREQUENCY TRANSDUCER
(BACK)

L L + −

L TERMINAL   + TERMINAL

L TERMINAL   − TERMINAL

Figure 9-1.Frequency Transducer Testing.

(2)Start the generator set and measure input voltage across terminals L-L with a multimeter.This voltage will normally be 120 volts AC for 60 Hz and 400 Hz sets and 24 volts AC for the 28 volt DC sets.

(3)Observing polarity, measure DC microampere current output on + and -

terminals be   of the frequency converter.This DC microampere a 0 to 200 microampere   current should

(1)   Connect a waveform (signal) generator across terminals L-L transducer.   on the frequency

(2)   Connect a resistance providing a total load of 1100 ± 27.5 ohms across

terminals + and- on the frequency transducer or connect transducer to frequency meter.

(3)   Observing polarity, measure DC microampere current output on + and -terminal of the frequency transducer.

Test frequency transducer as follows:

(4)   (a) 400 Hz frequency transducer.

1 Apply an AC input voltage between 114 and 126 volts to L1 and L2.

2 With an input frequency of 360 Hz output current should be 0.0 to 10 microamperes.

3 With an input frequency of 420 Hz output current should be 190-210 microamperes.

    <u>4</u>   Output current should vary linearly with an input fre-   between 0 and and
quency between 360

(b) 60

    <u>1</u>   Apply an AC input voltage between 114 and 126 volts to L1 and L2.

    <u>2</u>   With an input frequency of 55 Hz output current should be 0.0 to 10 microamperes.

    <u>3</u>   With a input frequency of 65 Hz output current should be 190-210 microampere.

    <u>4</u>   Output current should vary linearly between 0 and 210 microamperes with an input frequency between 55 and 65 Hz.

VDC frequency transducer.

(c) 28   Apply an AC input voltage between 19 and 28 volts.

    <u>1</u>   With an input frequency of 55 Hz output current should be 0-10 microamperes.

    <u>2</u>   With an input frequency of 85 Hz output current should be 190 to 210 microamperes.

    <u>3</u>   Output current should vary linearly between 0 and 210 microamperes with an input frequency between 55 and 85 Hz.

    <u>4</u>

c. <u>Calibration.</u>

(1) If the frequency transducer does not meet the testing requirements, recalibration of the transducer may be required.

(2) Remove access cover.

(3) The calibration screw offers the following adjustment ranges:

    ○   400 Hz Converter - 100 ± 20 microamperes (± 4 Hz) with 120 volt 400 Hz input.

    ○   60 Hz Converter- 100 ± 20 microamperes (± 1 Hz) with 120 volt 60 Hz input.

    ○   28 VDC Converter -100 ± 20 microamperes (± 1 Hz) with 24 volt 60 Hz input.

(4) Secure access cover.

d. <u>Replacement.</u>   If meter is defective, follow replacement procedures given in operator/organizational manual.

CHAPTER 10

GENERATOR SET TEST AND INSPECTION AFTER REPAIR OR OVERHAUL

Section I. GENERAL REQUIREMENTS

10-1. GENERAL. The activity performing the repair or overhaul is responsible for the performance of all applicable tests and inspections specified herein. Activities performing maintenance on any portion of the generator set must perform those tests and inspections required by the applicable component or system repair instructions.

Section II.   INSPECTION

10-2.   GENERATOR SET.

a.   Inspect entire set for loose, missing, or damaged parts.

b.   Check that each part is properly painted.

c.   Check that auxiliary fuel line and grounding rods are properly secured.

10-3.   ENGINE.   Inspect that proper lubricating oil is at the required level.

WIRING HARNESSES.   Inspect that all wiring harnesses are properly connected and that all connectors are tight. Check that all wires are secured away from moving parts which might damage wires.
10-4.

Section III. OPERATIONAL TEST

10-5. OPERATIONAL TESTS. Refer to Table 10-1 for operational testing of MEP-016B (60 Hz) generator set.   Refer to Table 10-2 for operational testing of the MEP-021B (400 Hz) generator set. Refer to Table 10-3 for operational test of MEP-026B (28 VDC) generator set.

Test	Test Method (MIL-STD-705)	Requirements
Open circuit saturation point at voltage and speed. rated voltage)	410.1	Maximum 0.25 ampere at rated curve (one

Test	Test Method MIL-STD-705)	Requirements
Rated load current saturation curve (one point at rated voltage)	413.1	Maximum 0.77 ampere at rated voltage and speed.
Power input	415.1	Minimum efficiency of 70 percent when measured after 1 hour stabilization at rated voltage, speed and load (0.8 pf) and 80 percent at 1.0 pf.

$$\text{Efficiency} = \frac{\text{Load (kw)} \times 100}{\text{Input Power (kw)}}$$

Test	Test Method MIL-STD-705)	Requirements
Overspeed	505.3	Perform at 110 percent of rated speed for 10 Minutes.
Voltage waveform (harmonic analysis)	601.4	Individual harmonic 3 percent maximum with all harmonics above 0.1 percent recorded at no load and rated load for 3 phase and single phase.
		200 percent of rated current for 10 seconds.
Short circuit	625.1	Maximum temperature rise for stator windings 75°C (135°F), at rated load at 125°F ambient.
Heat run (temperature rise)	680.1	
Humidity	711.1	Perform 614.1 in lieu of 608.1. The voltage regulation from no load to rated load and from rated load to no load shall not exceed 4 percent of rated voltage. There shall be no deterioration that affects performance of the generator assembly or insulation resistance less than 1 megohm.

Table 10-2. Test Schedule (MEP-021B/400Hz).

Test	Test Method (MIL-STD-705)	Requirements
Open circuit saturation point at	410.1 voltage and speed. rated voltage)	Maximum 0.50 ampere at rated curve (one voltage and speed. rated voltage)
Rated load current curve (one voltage and speed. point at rated voltage) Power input	413.1	Maximum 0.96 ampere at rated saturation Minimum efficiency of 75
	415.1	percent when measured after 1 Hour stabilization at rated voltage, speed and load (0.8 pf) and 80 percent at 1.0 pf.
		Efficiency= $\dfrac{\text{Load (kw) X 100}}{\text{Input Power (kw)}}$
Overspeed	505.3	Perform at 110 percent of rated speed for 10 minutes
Voltage waveform (harmonic analysis)	601.4	Individual harmonic 3 percent maximum with all harmonics above 0.1 percent Recorded at no load and rated load for 3 Phase and single phase.
Short circuit		200 percent of rated current
	625.1	for 10 seconds.
Heat run (temperature stator windings 75°C		Maximum temperature rise for rise)
	680.1	
	711.1	
		608.1. The voltage regulation from no load to rated load and from rated load to no load shall not exceed 4 percent of rated voltage. There shall be no deterioration that affects performance of the generator assembly or insulation resistance less than 1 megohm.

10-3

Table 10-3. Test Schedule (MEP-026B/28VDC).

Test	Test Method (MIL-STD-705)	Requirements
Open circuit saturation curve (one point at rated voltage)	410.1	Maximum 0.25 ampere at rated voltage and speed.
Rated load current saturation curve (one point at rated voltage)	413.1	Maximum 0.77 ampere at rated voltage and speed.
Power input Overspeed	415.1	Minimum efficiency of 72 percent when measured after 1 Hour stabilization at rated voltage, speed and load. $$\text{Efficiency} = \frac{\text{Load (kw)} \times 100}{\text{Input Power (kw)}}$$ Perform at 110 percent of rated speed for 10 minutes
Short circuit	505.3	200 percent of rated current for 10 seconds.
Heat run (temperature rise)	625.1	Maximum temperature rise for stator windings 75°C (135°F). At rated load, at 125°F ambient.
Humidity	680.1	Perform 614.1 in lieu of 608.1. The voltage regulation from no load to rated load and from rated load to no load shall not exceed 4 percent of rated voltage. mere shall be no deterioration that affects performance of the generator assembly or insulation resistance less than 1 megohm.
	711.1	

APPENDIX A

REFERENCES

1. **PAINTING:**

T.O.35-1-3   Painting and marking of USAF Aerospace Ground Equipment.

2. **RADIO SUPPRESSION**

MIL-STD-461   Radio Interference Suppression.

3. **MAINTENANANCE**

T.O.1-1-1	Cleaning *of* Aerospace Equipment.
T.O.1-1-2	Corrosion Control and Treatment for Aerospace Equipment.
T.O.35-1-11	Organization, Intermediate and Depot Level Maintenance for FSC 6115 Equipment.
T.O.35-1-12	Components and Procedures for Clening Aerospace Ground Equipment.
T.O.35-1-26	Repair/Replacement Criteria for FSC 6115 Aerospace Ground Equipment. USAF Equipment Registration Number Sytem Applicable to FSC 6115 Equipment.
T.O.35-1-524	Operator and organizational Maintenance Manual Electric Motor and Generator Repair.
TM 05926B/06509B-12/1	Electric Power Generation in the Field.
TM 5-764	Organizational, Intermediate (field) Direct Support and General Support and Depot
TM 5-766	Maintenance Repair Parts Lists.
TM 5-6115-615-24P (A)	
SL-4-05926B/ 06509B-24P/2 (MC)	Processing and Inspection *of* Aerospace Ground Equipment for Storage and Shipment.

4. **SHIPMENT AND STORAGE:**

T.O.35-1-4   Processing and Inspection of Non-Mounted, Non-Aircraft Gasoline and Diesel Engines for Storage and Shipment.

T.O.38-1-5   Procedures for Destruction of Equipment to Prevent Enemy Use.

5. **DESTRUCTION OF MATERIAL**

TM 750-244-3

6. **MAINTENANCE FORMS AND RECORDS:**

DA PAM 738-750 The Army Maintenance Management System.

DA PAM 310-1	Consolidated Index of Army Forms and Publications.
SL-1-2	Marine Corps Index of Authorized Publications for Equipment Support.
AFM 66-1	Air Force Maintenance Forms and Records.
TM 4700.15/1	Marine Corps Forms and Records Procedure.
NAVMC Form 10772	Recommendations for Changes and Improvements for Technical Publication.
AFTO Form 22	Recommendations for Changes and Improvements for Technical Publications.
DA Form 2028	Recommendations for Changes and Improvements for Technical Publications.

INDEX

Subject	Paragraph, Table Number
Battery Charger Stator and Air Scroll Back Plate	7-11
Battery Repair	4-2

C

Camshaft	7-16
Compression Test Faults T7-1	
Connecting Rod	7-15
Consumable Operating and Maintenance Supplies T2-2	
Control Box	2-8
Control Box Assembly	9-2
Crankcase 7-19	
Crankshaft and Generator Adapter 7-17	
Critical Torque Values................................. T1-2	
Current Transformer MEP-016B/MEP-021B (60/400 Hz) 5-10	
Cylinder and Piston 7-13	
Cylinder Head	7-6

D

Description 1-6	
Differences Between Models 1-8	

E

Engine Assembly	7-2
Engine Control Circuit Board 8-2	
Engine Inspection 10-3	
Engine Remove and Install	2-9
Exciter Stator Testing 5-7	

F

Fabricated Tools and Equipment 2-3	
Flywheel and Engine Fan 7-10	
Frame, Removal, Repair, and Installation 3-2	
Frequency Meter	9-3
Frequency Transducer 9-4	
Fuel Injection Pump	6-3
Fuel Injector 6-4	
Fuel Tank	6-2

Paragraph, Ta-
ble Number

G

Gear Cover and Seal 7-12
General Maintenance 2-6
General Maintenance   Requirements ....
Generator 2-10
Generator   Assembly 5-2
Generator   Bearing 5-8
Generator   Fan 5-3
Generator   Set 10-2

L

Levels of   Lifters and Push Rod Tubes   1-5
         Maintenance Accomplishment   7-9
...... Limited Applicability
                                 1-2

M

Maintenance Forms and Records   1-3
Malfunctions Not Corrected by Use of the Troubleshooting Table   2-5

O

Oil Cooler 7-20
Oil Filter Adapter 7-18
Oil Pan 7-3
Oil Pump 7-14
Operational Tests 10-5

Q

Q1Test Chart T5-19
Q2 Test Chart T5-2

R

Repair and Replacement Standards T1-1
Repair Parts 2-1
Reporting of Errors 1-4
Rocker Arms and Push Rods 7-8
Rotating Rectifiers (Diodes) 5-5
Rotor, Removal, Testing, and Installation 5-4

Subject

**Paragraph, Table
Number**

S

Scope 1-1
Skid Base. 3-3
Special Tools, Test and Support Equipment T2-1
Standard Torque Values T1-3
Starter Assembly   7-4
Starter Solenoid 7-5 5-6
Stator Testing . . . .

T

Tabulated Data 1-7
Test Conditions T5-3
Test Schedule T10-1
Tools & Equipment 2-2
Troubleshooting. T2-3

V

Voltage Regulator   5-9
Valves . . . .

W

WiringHarness 10-4